Volitional Attention Training

Neural Plasticity and Combative Application

Kevin J Mills

Published by Kevin J Mills. PO Box 107, Crediton, Devon, EX17 9AR

For enquiries e mail : spire0951@aol.om

ISBN: **0957604726**
ISBN-13: **978-0957604728**

DEDICATION

To all who seek knowledge with an open mind.

To my soul mate Jenni and to my ever growing family, Leah, Scott, Deezyl and Aaliyah.

CONTENTS

DEDICATION

FOREWORD

ACKNOWLEDGMENTS

INTRODUCTION

1   TO THINK WHAT HAS TO BE THOUGHT   1

2   PAY ATTENTION YOUR GUT IS TALKING   16

Similarity: Appearance Equals Reality

Framing: It's my Choice

Anchoring and Adjustments: So Come Up With Random Figure

Status Quo

Sunk Cost

Confirmation

Cognitive Over Confidence

Risk Aversion

Selective Perception

Closure

Common Fate

Instincts and Your Intuitive System One

Priming

3   **MEASURING LIFE BY THE NUMBER OF**   40
           BREATHS

OODA Loop

Breathing

Sensory Acuity

Training our Attention

Methods of Practice, Pranayama.

Tactical Breathing

Brain States While Being Mindful

Gamma Waves

4   **THE BODY SEEKS SYMMETRY**   58

Left Vs Right Hand

Learned Maladaptive Behaviour

Natural Movement

Hemiplegic Gait

Diplegic Gait

Neuropathic Gait

Myopathic Gait

Parkinsonian Gait

Choreiform Gait

Ataxic Gait

Sensory Gait

Activity Dependent Cortical Reorganization

Symmetry of Posture

Symmetrical Mirror Postures

5   THE WARRIOR MIND   90

Queen Boudicca

"Mad" Ann Bailey

Warrior Tribes

Becoming a Warrior

The Language of the Warrior

Bravery

Honour

Magnanimity

Valour

Courage

Virtues

Shiba Yoshimasa (1349 – 1410)

Ichijo Kaneyoshi (1402 – 1481)

Nakae Toju (1608 -1648)

Kumazawa Banzan (1619 – 1691)

Yamamoto Tsunetomo

Indians

6   VIOLENCE IT'S NATURAL LET IT BE   121

Category 1 Social Violence

Category 2 A Social Violence

Cultural Excuses

Evolution

Our Ancestors

Plant Domestication

Crime and Violence

Tochelson and Samenow (2015)

The Construct We Call Mind

Genetics

7   NEUROMUSCULAR PROGRAMMING   153

Somatosensory System

Proprioception

Constraint Induced Movement Therapy

Learned non – use, How Habits Form

Learned Bad Behaviour

8   STARTLE REFLEX, I DIDN'T JUMP!   174

Hick's Law

Power Law of Practice

Reaction Time Responses

Body Reflex

What Should you Teach

Spinal Reflex

Brain Stem Control

Higher Cortical Control

9   PERFORMANCE UNDER STRESS   196

Stress – Anxiety – Fear

Internal Physiological Responses

Physical Responses and Disruptions

Death Grip

Freezing

Perceptual Narrowing

Audio Exclusion

Motor Control Interruption

Internal Psychological Responses

Awareness Failure

Decision Errors

The Thermoregulation System

Managing High State Emotional Arousal

1. Conceptualisation

2. Skill Acquisition

3. Applications

10   THE BULLY, AGGRESSIVE AND IN CONFLICT   218

Why Are Children Bullies?

Primate Evidence

Cultural Reasons For Aggressive Bullying

The Terrible Twos

Born And Not Made

What's Happening Within The Brain?

A Plan of Action

Martial Arts

Be Mindful All The Time!

Cyber Bullying

Anti Abduction Skills

The Bully Who Knows No Right From Wrong

Fear

REFERENCES

ABOUT THE AUTHOR

FOREWORD

"I have been involved in violence for many years from everyday confrontations and scuffles to lethal force encounters and I can vouch for the content and relevance of Kevin's book to that 'dark realm'. Violence is both art and science and Kevin clearly explains this concept in a comprehensive, thoroughly well researched and written book that is a shining light into that 'dark realm'. This book is required reading for all serious students of self defence/combat and considered as a classic of its genre.

Mo Teague

ACKNOWLEDGMENTS

What else is there but the experience?

There really is only one who has been there beside me throughout my martial arts journey, my soul mate and devoted wife, mother and grandmother Jenni.

No words can ever truly explain the love and support that you have provided. Your time in patiently editing my sometimes illegible words and for working out what I want to say.
Thank you.

INTRODUCTION

After the completion of my first book "The Secret Science of Modern Martial Arts" I found myself looking at a massive amount of material that I didn't use! The next step was an easy one; I needed to put more information down. Since 2005 I have been constantly working on a method of training that would be focused on one very particular element of violence, understanding and training a mentality. From this came Volitional Attention Training.

V. A. Training

Volitional Attention Training was designed to specifically develop a reality based, combative application of both physical and mental training within one program. The training enables individuals to manage and cope with high-end emotionally stressed deadly force encounters, when your life is on the line.

This is a "fight for your life" method designed to deal with A Social violence, it is not self-defence, it is SELF PRESERVATION.

V.A.Training uses the latest research into the ability of humans to use both physical and psychological applications during violent encounters. Science has allowed humans to understand how the brain and the mind work and how to create plasticity within the brain. The program uses this understanding and knowledge to ensure that the highest levels of performance are achieved.

- Volitional and Mindful
- Latest method of application
- Target specific
- Blade psychology
- Combat applicable
- Stimulus training
- Biomechanical Movement

This method of training taps into our willful mind, it will teach you to access the brain's processing mechanisms via the mode of attention, this will enable intuitive and natural responses to high emotional situations. Volitional Attention Training will guide you in creating your own mentality, which in turn will give you the tools to engage aggressive predators and to access the most natural synchronised movement necessary to deal with A Social violence. Not only will this training teach you the physical movement and techniques required to move to an advanced level of combat, it will also teach the psychology required and how to develop very powerful methods of behaviour. You will learn why we do the things we do and how to control and use them to your advantage. When all else fails, and you are fighting for your life Volitional Attention Training will provide you with the psychological and physical skills necessary to survive, to protect your love ones and ultimately your life.

Whatever you need to do to make an aggressive violent predator non-functional is contained within this training method. This is not a system; but methods of advanced combative applications in the worst of situations. It has been designed to provide knowledge, understanding and ability. One of the greatest benefits is that Volitional Attention Training can be bolted onto any existing fighting style or martial art, it can be learnt as a stand alone method or in conjunction with what you already train in.

Volitional Attention Training uses the latest and most up to date research into the workings of the human brain and why our mind thinks the way it does. It will help you develop and create effective and efficient movement, drawing on the field of neural plasticity to ensure spontaneity and intuitive responses. The aim of Volitional Attention Training is nothing short of what is required to survive both mentally and physically when faced with violent and aggressive behaviour and when ultimately your life is on the line.

The Science and Psychology

This book will give you an insight into the supporting architecture of V.A.Training, it is very important that as an individual in your own right, you are capable of first understanding the reasons why the training has been developed in the way it has. This is not a monkey see monkey do training program, it's completely the opposite, and you need to know why you are training in a specific method. In knowing the reasons why it will give you;

- Confidence in your skills and ability
- Change your behaviour and your attitude
- Motivate you to continually question and push your own boundaries
- Provide you with the knowledge and understanding that will enable you to share what you know
- Move in the most effective and efficient way

You will discover some myth busting knowledge regarding violence and how to manage and cope with the worst of situations. Above all what this information will do, is expand your mind and the way you see self-preservation.

What should I do with all the knowledge that I have learnt over many years of experience after experience?

Share it, with whoever will listen.

1 TO THINK WHAT HAS TO BE THOUGHT

What is attention or mental force, how does it create neural activity and what are its benefits? "The task is not so much to see what none have yet seen, but to think what nobody has yet thought, about that which everybody sees" Schrodinger, E.

The hardest attribute to relay to any student of the martial arts is not found in the physical realm, but rather the mindful application of "mental force" Schwartz and Begley (2002), which all humans are capable of harnessing. This mental force comes in all manner of forms and descriptions, indominatable spirit, warrior mind and attitude, are all examples of these. However, a more important question should be, how is this state of mind achieved, and what processes occur within the mind and the brain? To help answer these questions will require an understanding of an area of science and physiology not often explored, namely that of mental thought processes that create will power or volitional effort. "Volitional Effort" is effort of attention, the function of the effort is to keep affirming and adopting a thought, that if left to itself would slip away, effort of attention is thus the essential phenomenon of will" James, W. (1890). There are a few individuals in whom this type of mental force seems to be present in abundance, then there are those, and these are in the majority, that do not possess this mental force in any way. They have been molded over their lifespan through behaviour and an exposure to either a physical experience of violence or a thought process that never required them to engage in what could be termed aggressive thoughts or the ability to use will power to overcome a stressful situation.

In essence, psychological skills are required to help support physical skills. Mental toughness, mental force and attitude of mind need to be explored and defined. This involves two specific areas: - 1, the actual processes that are taking place within the brain; and 2, the mind's ability to channel attention and mental force. There are individuals that seem to possess these abilities in abundance, if this is the case, important questions would be, how did this attitude of mind develop and is this the product of behaviour and social identity created by circumstance? Children, directly as a result of significant caregiver roles within the family unit, often inherit behaviour habits. Having a parent with aggressive tendencies could lead to transference of aggressive behaviour to any child, male or female. Equally, an over aggressive caregiver could cause a complete lack of self esteem, leading to withdrawal of that individual, who also lacks the ability of mind to be confident, and bring forth the mental will power required to create mental force. If behaviour habits are so important, what constructs and processes are affected within the brain?

Evolution also plays a part in our understanding of mental force and the benefits derived from possessing it, with a direct link to Darwin and the survival of the fittest. Imagine a history where humans did not possess these types of abilities, would we have ever dragged ourselves out of the primeval world that we occupied? There are mental processes that have to be overcome in order for any individual to live a life, to find a mate, reproduce, to survive! To enable this process, the mind as well as the physical body, has to be mentally healthy and fit.

As humans, we are constantly under threat from our mind's activity, we therefore have to understand what is happening when certain moods take over the dominance of our minds, or when we create thoughts that are not congruent with our mental direction. Maintaining the physical body has to form part of this process, ensuring that the body is kept in a state of physical wellbeing will result in a positive attitude, if an individual suffers from a physical impairment, is obese, sleep deprived, lacks nutritional balance, inputs substances into the body (drugs), then the consequence of this is a human organism that is not in balance, the body and mind do not work as one. If the mind was mentally tough and capable of survival and the body was not, it would not take long for one to adversely affect the other, or vice-versa. Therefore physical conditioning should be equally as important as mental conditioning.

Bringing these ideas into attention earlier in this discourse creates an understanding that attention has to be thought about. A stimulus input into the brain creates a mechanism of mental processes, that in turn leads to an amount of mental attention being applied to that stimulus, how long attention is maintained will depend upon the amount of mental focus that the individual is capable of bringing to bear upon the stimulus. A stimulus that brings forth an episodic memory will also bring with it the ability for the mind to pay more detailed attention to that particular thought. Episodic memories are those that are encoded into the mind, through an emotional experience, these experiences are capable of coding in the time, place, feelings and details of the event, they are far more real to the mind than attempting to memorize an event to which you are just a passive observer.

Semantic memory is generally concerned with knowledge of the world that we live in, there is a difference between, knowledge that is factual and personal experiences that have encoded knowledge and understanding with a greater grounding and meaning. Both semantic and episodic memory deals with long term, rather than short-term memory, a key difference is that episodic memories encode the actual acquisition experience and the context in which the memory occurred. For any combative or martial art technique to become efficient and effective, the coding process will need to support the intended action, i.e. self-defence techniques will have to become linked to procedural memory. Declarative memory deals with facts and data gained from learning, "declarative memory serves to "chunk" or "bind" together the converging processing outcomes reflecting the learning event, providing a solution to the "binding problem" for memory, Cohen, N. Poldrack, R. Eichenbaum (1975). The sea is wet and the sun is hot are examples of long-term declarative memories.

Procedural memory is concerned with long-term memory including complex motor skills. These skills are first coded into the mind and over time become second nature; you do not have to use a cognitive thought process to access the skills. Playing a musical instrument, driving a car, or combative, martial art techniques, are all examples of procedural memory, "procedural memory enables organisms to retain learned connections between stimuli and responses, including those involving complex stimulus patterns and response chains, and to respond adaptively to the environment" Tulving (1985). There is no defined limit to long-term memory and providing that the correct coding procedure occurs, complex motor skills that involve combative and martial art techniques can be built up. Continued repetition of these movements will lead to a stable procedural memory, which ultimately leads to spontaneous movement. This is arguably the aim of any person engaged in this type of activity. It is important to remember here that any human movement can be learnt in a manner that is not congruent with natural movement, it is maladaptive.

Continual repetition of techniques that do not follow this premise will eventually cause damage to the organism. Occupations that involve high stress and the potential for deadly force encounters are particularly exposed to incorrect episodic memory input, and again, continued exposure to this type of maladaptive behaviour, could have disastrous consequences, "in the blink of an eye, the officer snatched the gun away, shocking the gunman with his speed and finesse. No doubt this criminal was surprised and confused even more when the officer handed the gun back to him, just as he had practiced hundreds of times before" Grossman, D. (2008). This is a good example of incorrect coding of a maladaptive procedural memory. The officer involved continually practiced this disarm, until he had coded it into his mind, in doing so creating a spontaneous response, it had become second nature to him, I term this "negative loop coding" (NLC) which should be avoided for obvious reasons.

The disarm in itself was never the problem, in fact over time several episodic events could have occurred in this officer's life, for example, he may have already been associated with lethal force encounters, he may have had colleagues die in the line of duty, any of these high emotional states would have led to an episodic memory. Once the officer had started to pay attention to this training loop and began to practice the disarm in all sorts of situations, both at work and at home, he had started to encode procedural memory, the only problem with the training was the handing back of the weapon! To do it again and again, and again! A key point in this behavioural pattern is volition used to pay attention. Once attention on the training pattern had begun, his brain would have been firing neurons at a fast pace to start the encoding, drawing with it greater amounts of mental force, enabling focused thoughts on the reasons for the practice, in other words the officer was undertaking mindful attention.

Grossman (2004) provided further evidence for NLC, he explains about the training methods of FBI agents and their procedures during pistol training, "Officers were drilled on the firing range to draw, fire two shots, and then re-holster. While this was considered good training, it was subsequently discovered in real shootings that officers were firing two shots and re-holstering, even when the bad guy was still standing and presenting a real threat!" To understand the way in which the brain works and how this affects the mind's ability to make decisions, and to draw from hard-wired memory of physical movement is imperative. NLC has to be avoided at all cost; no distinction should be made between an officer or military personnel entering a deadly force encounter, or a door supervisor or regular citizen going about their daily lives and having to confront the aggressive thug!

Any aggressive encounter brings with it the potential for real harm to occur. Life or death could hang on the split second that it takes for the brain to produce a response, generated through the process of episodic memory, embedded in procedural memory, that create spontaneous reactions that are maladaptive, rather than adaptive for survival. To over look this, to not understand the causal effect of NLC, is to put at risk any individual who may be relying on your knowledge to protect them. For the combative, military, police, or martial art students, the context of the training and ultimately the encoding procedure is critical.

Experiments were conducted by Passingham, R. and colleagues of Oxford Universities Institute of Neurolology, using Positron Emission Tomography (PET) scans to test the firing regions of the brain, both before and after a new complex motor skill was being encoded. The subject was given a task of remembering a correct 8-digit sequence on a keypad, obtaining the correct key order by trial and error. Every time he pushed the correct key a confirming sound would happen. The subject had to pay attention to the sequence of keys as well as having to control his motor skills. The regions of the brain that were particularly active were the pre frontal cortex, parietal cortex, anterior singulate; chordate and cerebellum, all are used to control planning, thinking and moving. After a period of time the subject was able to remember the sequence and was instructed to continue entering the sequence until it could be done without any thought or errors.

When at this stage a PET scan was repeated, it was discovered that all the previous areas of activity had been shut off; all that was now active was the motor cortex region of the brain.

This experiment and the processes of its encoding, can be linked to combative and martial art techniques. There are varying degrees of techniques that are taught today, ranging from very simple movement to complex, linked movements that can involve detailed motor cortex skills. Any new individual undertaking such processes will be using their mind in the same manner as the research subject above. Their brain's neurons will be firing away in all the same regions as above. The caveat here is that they will also have to focus attention to the task in hand. If the individual that is attempting to learn such skills, has no motivation to learn them and is not applying mindful attention, then the skills are not going to be encoded into their memory centres, other than the short-term memory. If this is the case, then as the description implies, they are not going to remember many of the techniques taught, especially when they are required at the most critical time, when a high state of emotion or arousal has occurred. In this situation the individual will not be able to create either episodic memories or procedural memories and therefore will not be able to encode the techniques into subconscious, spontaneous movement.

Interestingly, in the above research experiment, when the subject was asked to re-focus their mindful attention back onto pressing the correct keys again, which he could do without thought, the neurons within the brain once more ignited within the regions previously noted, providing evidence that switching back on the thought of attention through volition, had a causal effect on the brain itself, clear evidence that our own thoughts have a physical effect on the mechanisms of the body.

Episodic memory generation should not be taken lightly given the nature of violence, as any individual experiencing violence will in all likelihood experience an episodic event, one which will stay with them for a very long time. Post Traumatic Stress Disorder (PTSD) is a type of severe anxiety disorder, which usually develops after an intense, extremely traumatic event, called a "stressor" This stressor could easily be an experience of violence, especially anti social violence or a deadly force encounter. Individuals who are exposed to traumatic events experience psychological stress; this can, if the stimulus that has created the event is intense enough, result in the long-term creation of certain behaviours, for example, agoraphobia, the inability to go outside. The causal events of trauma for an individual encompass a varied range of experiences such as road traffic accidents, violent crime, death of a loved one, rape or aggressive violence. At a social level, events such as war, conflict, natural disasters and acts of terrorism can affect large amounts of people exposing them to trauma, which in turn leads to psychological stress. Individuals will have different levels of stress responses to trauma, dependent upon their psychological character, they will include acute fear, grief, behaviour changes and changed beliefs about the World and society. Therefore it is easy to create a link between attention, episodic memories and PTSD, the one thing that they all have in common, is their ability to focus the mind, to take up huge amounts of mental energy, enabling the mind to focus it's attention on thoughts that are kept in the forefront, from a bi-polar effect on the mind. These thoughts can either enhance, or degrade an individual's performance under stress; it is for these reasons that understanding the mind's ability to create positive re-enforcing thoughts rather than negative ones, is so important. You can have the most effective self-defence techniques that have been developed today, but if your attention is not completely in line with your intentions then you will be defeated. There is a saying I once read, "when skill ends, the only thing left is your mind" if your mind is not in the fight then you have lost before you have even began.

This leads to an understanding that the power of thought and the ability to pay attention has a causal effect on behaviour and our mental states. The type of thought is also associated to different types of stimulus, for example, unconditioned stimulus, which is linked to an unconditioned response. This was initially brought to attention by the Russian psychologist Ian Pavlov and his experiments with dogs. Pavlov (1902) started from the idea that there are some things that a dog does not need to learn. For example, "dogs don't learn to salivate whenever they see food. This reflex is 'hard wired' into the dog. In behaviorist terms, it is an unconditioned response (i.e. a stimulus-response connection that required no learning)." Simply psychology (2007). This type of learning is know as 'classical conditioning" what Pavlov observed was that his dogs salivated on the presentation of food, this he termed an unconditioned response to an unconditioned stimulus. By chance, he then discovered that the dogs started to salivate prior to the presentation of food; they began to salivate when the research assistant appeared. This led him to an understanding that the dogs had learnt a behaviour, as they were not salivating on seeing the assistant to begin with. He then tested this theory by introducing a bell when the food was presented, this he called a neutral stimulus.

After a period of time he then presented the bell stimulus without food, observing that the dogs salivated and confirming that they were able to learn a new behaviour. The neutral stimulus had developed into a conditioned stimulus and the salivation was a conditioned response. This type of conditioning occurs throughout our daily life, classical conditioning will affect us in either a positive or a negative way. PTSD is an example of negative maladaptive behaviour in its most extreme; a traumatic event has created anxiety. The event coding could have started from a simple stimulus such as a light flashing! Imagine, you are a firefighter, the call to action is signaled by the flashing of a red light, both in the station and on your way to a shout. Over a great number of events nothing has happened that created a traumatic memory, which encoded itself via episodic memory. Then one day you are feeling a little vulnerable and you are called to a fatality involving a person that has been killed in an unpleasant way and the one thing that you remember before this is a flashing red light! This light has now become your "stressor" and every time you see one it triggers the memory (episodic) of that nasty event.

Your mind fixates on this, your attention is drawn back to that day, a conditioned response has been created, and anxiety and stress are the result of this attention. Once this has been encoded into our brain, reversing such emotional responses is very hard. It is therefore very important to be able to protect one's own psyche when entering into any high emotional situation. The reverse is also true; associating pleasant emotions with a stimulus can just as easily result. A place or a song that brings forward happy memories can also create conditioned responses, such as a smile or laughter.

Another method of learning is operant conditioning; this type of learning involves either positive or negative reinforcement. This is where behaviour is rewarded. This type of learning also has to be considered carefully, as reinforcing a maladaptive behaviour could be disastrous. I have personally witnessed students being told how great they are, when from my context of understanding there was never good movement or substance shown. This type of continued praise has to be considered carefully, as on one side of the coin it's building confidence and self esteem and on the other it's creating maladaptive behaviour. Consider the example above of the FBI agents being instructed that to shoot twice and then re-holster was effective and professional marksmanship! At the time of instruction, those performing well would have certainly been rewarded, both by the training staff and colleagues around them, reinforcing the learning and mechanical processes of shoot twice and re-holster.

It is completely understandable that at the time that this type of operant learning was taking place, the individuals involved would have taken the instruction without thought of the consequences that the NLC would create, "Tragedy born of ignorance usually winds up becoming the event to turn the tide towards the quest for knowledge, so that similar tragedies will never be experienced again" Murray, (2004).

Our understanding of the mind, its learning patterns and more importantly how the body responds automatically has drastically improved over the years and it's this continued progress that needs to be understood to ensure that instructors responsible for teaching individuals any type of discipline that could involve a high risk of personal injury or deadly force encounter, are at the cutting edge of the thought process. Simulation of experiences and stimulus input is therefore very necessary, if reality is the goal of the discipline being taught.

Martial arts or combative instruction is no different from police officer training or military training, if the student is continually praised for their movement, and this movement or the psychology involved is built upon maladaptive behaviour, then there could be serious consequences. Obtaining knowledge and understanding of tactics that are up to date in today's environment is not an easy task, it is certainly one that takes a certain moral compass to evaluate and change, not relying on ego to drive direction continually forward without reflection. I like the four A's, Assess, Analyse, Adapt and Adjust.

Assessment of a technique is the start of this thought process, here I will give a brief overview, later I will discuss this in more detail. For any individual to be able to assess a technique they have to have a certain amount of knowledge to do this. They will also need some sort of process to go through. Analyszation, requires taking the detail and deconstructing it to its basic level, at the same time remembering the context of the technique. Once this has been completed you are in a position to adapt the technique and if necessary make adjustments. Once this process has been completed the technique is ready to be re-encoded back into your mental thought processes.

What this evidence provides is confirmation that what we pay attention to, how we think and what we think about, can be critical. If we are to learn from past failures and in some cases very tragic losses, then we have to think about what often we do not want to, only in this manner do we stand a chance of being able to develop mental force, attitude, a warrior sprit, it's up to you what you call it, but this has to be paid attention to. It is clear that attention plays a key role in the activity within our brains

There is an element of mindful control that has to happen, in order for attention to be exactly that "attention" The brain has to fire its neurons, creating action potentials in the particular part of the brain that is receiving the stimulus; these mechanisms are focused on by the brain and in turn create attention. The amount of sensory input that the brain receives every second of every day is staggering. We see, hear, smell, touch and feel, yet we do not pay attention, until something draws our attention towards a stimulus event " attention defines the mental ability to select stimuli, responses, memories or thoughts that are behaviorally relevant, amongst the many others that are behaviorally irrelevant" Corbetta, (1998). What is relevant will wholly depend upon the current situation and incoming stimuli, if this happens to be a high stress and emotional one, then attention will be directed in such a way that the bodily responses are congruent with prior thought processes. If there is no link to positive mental processes of mental force then a degrading of attention may occur. While all this is occurring the body's internal control mechanisms are also working at full tilt, providing even further stimulus input that the brain is having to deal with, without any cognitive awareness.

Stimuli from our external senses are not the only way in which attention can be created, close your eyes and imagine something that brings to your mind a vivid picture in your minds eye, a bright red rose, waves from the sea crashing upon a sandy beach, or the face of a loved one. Each time focus is attended to, through conscious will power, attention can be maintained and your neural network jolts into life. Meditation uses just these processes to produce physical changes within the body. For years, before the invention of machines that could measure and record brain activity such as Functional Magnetic Resonance Imaging (FMRI), Computerized Tomography (CT) or Positron Emission Tomography (PET), meditation was viewed as some kind of mystical activity, with no real substance or evidence of the processes that were taking part in the brain.

Now we have evidence of the regions of the brain being engaged, when the mind takes control of attention and focuses on internal or external experiences "several studies have investigated the functional anatomy of covert visual orienting to simple unstructured peripheral stimuli. These studies have shown that a specific set of frontal parietal regions are consistently recruited during visual orienting" Corbetta (1998). Covert and overt visual orienting according to Corbetta are two distinct ways in which we explore our visual environment, by saccadic eye movements that happen naturally "overt" or by volitional attention or reflexively when a stimulus appears in our visual field " covert", the latter being the process when a sudden unexpected threat arises. A simple example of this could be an incoming punch; attention has to occur focusing mental force to deal with this threat.

The brain has been compared to a super computer and is by far the most complicated organism that we are currently aware of, the act of "attention" requires volitional effort, will power, and the brain using different regions to process received stimulus. Different areas are allocated to interpret and fire the neurons depending on the action that is being supported, what area of the brain would be used for motor control, or for creating a certain mentality to cope with a stressor for example? Can the brain be exercised just like the body and in doing so grow and develop in the same manner? What benefits would this provide to an individual learning a combative or martial art? When bodily and environmental awareness is the focus of attention, the parietal cortex is the region recruited within the brain to manage this type of stimulus input. For the formation of habit and coordination of movement the region recruited is the underlying cerebellum and basal ganglia.

To enable attention to be paid to motor responses that are required to support actions necessary to first recognise an intended threat and to then coordinate a response, multiple regions of the brain will be used to support this action. Increasing activity in these regions has been evidenced by FMRI "attention then is not some fuzzy ethereal concept, it acts back on the physical structure and activity of the brain" Schwartz and Begley (2002). Remember attention is a volitional process of will, therefore when we pay particular attention to a stimulus we increase the activity in the region of the brain responsible for this action, ' attention can sculpt brain activity by turning up or down the rate that particular sets of synapses fire" Robertson, I. cited in Schwartz and Begley (2002). This continued repetition of brain activity as we have learnt, reinforces the physical connections within the brain "it follows that attention is an important ingredient for neural plasticity" Schwartz and Begley (2002). Coming back to the title of this chapter, thinking is attention and what has to be thought when entering any type of high stress, emotional situation is imperative to survival.

Volitional attention, paying attention to thought, literally changes the brain. To think about negative outcomes will only increase the possibility of it happening, to think about a positive outcome will reinforce the end result. Once we understand this clear process we need to explore what, in a combative/martial context, should be thought about? We also need to consider how to teach an individual to think, focusing as much on this aspect as we focus on teaching skill. This in itself presents several problems, not least of which is the initial goal of a student entering a school. In the vast majority of cases, they will never experience deadly force encounters, let alone a brawl in the street or local bar, they will not understand the need for them to learn how to think, as we all know how to do that right! If "reality" is your buzzword, psychological preparation and tactics should already be an integral part of your training syllabi, but remember that ego can sometimes distort your thoughts.

If you haven't already, now is the time to fully think about your own mental capabilities. Are you capable of doing what has to be done to survive? Are you capable of going towards the threat rather than away? Contrary to what most think, the facts tell us that in the majority of high emotional violent encounters, the average human will not engage in such activity, they would preferably to get away from danger rather than go towards it, they are also not likely to want to kill a member of their own species.

In the book On Killing Grossman (1995) kill rates during successive wars over the centuries have progressively increased. Grossman presents evidence that during early civil wars in America, the kill rate was extremely low, between 80 and 85% of soldiers engaged in mortal combat would not aim to kill their enemy. Once the military powers had discovered this psychological barrier, which could be referred to the lack of a mentality to kill, procedures were taken to correct this "the training methods that increased the firing rates from 15 percent to 90 percent are referred to as 'programming' or 'conditioning' by some of the veterans I have interviewed, and they do appear to represent a form of classical and operant conditioning (a la Pavlov's dog and B.F. Skinner's rates). What is interesting here is that the innate seemingly inbuilt responses of the mind to killing have been overridden, by the programming of the brain through these two methods of powerful exposure to a stimulus. Evolution does have an answer to why "we will" kill our own and this I will look at later, for the moment however let's concentrate on "paying attention" to the actual thoughts.

You may have psychological barriers in place to prevent the clear thoughts that are necessary to move forward towards a threat. If your goal is effective, efficient and decisive responses to violent stimuli, then these hurdles have to be overcome. If you are in an occupation that regularly exposes you to high risk, deadly force encounters, then knowing the boundaries, both professionally and personally, need to be very clear. If you are Mr. Joe Average, then you also have to have very clear reasons for your actions. If you are a trained martial artist and you disable a person for life for spilling your drink, then you deserve every second of incarceration that you receive.

Whichever situation applies to you, you must first have thought about "paid attention to" the act of violence! If in your lifetime you only think about carrying out the act of killing in a justified situation, then at least you will have thought about it. To never think about it results in those neural pathways never ever being fired. To enable the majority of humans to enter into the theatre of violence, this also has to be thought about. Not only do you have to pay attention to the act itself, you also have to have thought about your technique, it could be shooting, knife-work, striking, it matters not, as we have read and understood above, thinking the actions and movement is just as important, if not more so, than the skill required.

Understanding the above provides a certain moral code, that it is actually ok to have considered, what a great many will shy away from. Remember the only difference between an officer of the law, a military solider or Mr. Joe Average, (martial artist) is the likely percentage of high emotional encounters being stacked against the martial artist or combative student. However the learning processes should, in either case, be as close to simulated truth of reality as one can get.

Using the mind is equally important, no matter which side of the fence you are on. To say that there is no need to pay attention to deadly force encounters, is not to understand how the mind and brain work, or the mechanics of the human body and ultimately to put at risk the people who have entrusted you with instructing them. Mentality is a state of mind and not a physical causation due to the mechanical workings of the brain, there is a difference, something that is real. Neural scientists have been able to map the brain; they have identified which regions of the brain are activated when we experience pain, or an emotion like anger and joy. They have however, to date, not been able to create the link between the mechanical processes and the conscious experience "Neural scientists have successfully identified the neural correlates of pain, depression, and anxiety, none of which amount to a full explanation of the mental experience that neural activity underlies, the explanatory gap has never been bridged and the in-escapable reason is this. A neural state is not a mental state, the mind is not the brain, although it depends on the material brain for its existence as far as we know" Schwartz and Begley (2002).

To create a mentality therefore has to be thought about, it does not just happen! Even those that seem to possess this super human quality, will have developed it throughout their life, through mimicked behaviour patterns and subjective passive learning. This leads us to one conclusion, if mentality is not the brain, then it is a product of the mind, created by the mind out of attention and mindful focus and as the brain has the capacity to rewire itself, "neural plasticity", then anyone can, by very defined process create a state of mind. You can train the mind to affect the brain and in doing so, enhance just about any activity that you choose to undertake.

2 PAY ATTENTION YOUR GUT IS TALKING

Consider this very straightforward question, what is the difference between the mind and the brain? Which one of these is responsible for that feeling that, something is just not right? You know that thing that we call gut, our instincts, those that we believe have protected us at some time in our long forgotten history, allowing us to survive the predators of that time.

To enable an individual to commit to training, they have to be secure in the knowledge that what they are about to undertake will provide them with the desired outcome and in today's environment, that is coping with the predators that walk our street, the thugs and petty criminals being in the wrong place at the wrong time or the professional that has to deal with these feelings on a day to day, month to month basis. I remember talking with a US ranger, retired special forces guy, you know the type of person that films are made of, one that has at every turn in the road stepped forward to go where most fear to tread, I remember clearly his words "I ignored my instincts nine times and each time, I was either shot or stabbed". Any training that is maladaptive or does not contain procedures that tap into this long forgotten sixth sense may ultimately fall short. If your training includes an understanding of instincts, what they are, how to recognise them, what they are not, then you are again on the path to a personal understanding, that uses the most powerful tool in our armory, that which has been responsible for dragging us along that evolutionary road to today's modern man, the human brain and the mind that lies within.

To start this process we first have to go way back, to the first society that proposed the hypothesis of two brains. The first people to propose this were the ancient Greeks. It's obviously not two brains just two systems and for a change they are named system one and system two, they are also known as 'Dual Processing systems'. In his book the Science of Fear Gardener (2008) used the term 'head' and 'gut' to explore the thought processes that are used by the two systems, as they are distinctly different. These terms are very appropriate to this discourse and so I will use them here as well. System two is labeled, "Head" and is responsible for reason, this is our conscious mind the one that we engage when we consider a situation, it works at a much slower pace than gut, taking its time to calculate, consider, working with logic and what it believes is the correct thought or answer.

System one is labeled "Gut", this is our subconscious mind at work, which is directly linked to our evolutionary past and is responsible for our survival and development to this day. Unlike system two, system one is super quick, it creates thought and transfers this to our conscious mind in a split second, gut has no time for the slow processes that Head has to work with. Gut is the source for the feelings of fear, unease, it makes the hairs stand up on the back of your neck it triggers your fight or flight response.

The US ranger story above is an example of system one sending signals from the subconscious mind, warning you that there is something not quite right with the situation, Head then gets involved and considers the situation, allowing time for the Head to over-rule the Gut and in the case of my friend above nearly costing him his life. "The idea that System 1 cognition is ancient and System 2 cognition is modern, in evolutionary terms, is a recurring theme in dual-process theories. This is often linked to the assertion that while System 1 cognition is shared with other animals, System 2 cognition is uniquely human. The last idea arises from its association with uniquely human processes such as language and reflective consciousness and the apparent ability to perform cognitive acts (such as hypothetical simulation of future and counterfactual possibilities) that are assumed to be beyond animals" Evans J (2006). One thing is very clear, system one is linked to thoughts that are produced almost instantly and the evidence suggests that this system is part of the mechanism that our long lost stone age ancestors used to alert them to impending danger, or when they were the main course on the menu. This system would have been selected over and above system two as an evolutionary adaptation, to enhance survival. Now, modern man is the safest he has ever been and does not usually find himself being hunted for dinner, he now has more distractions for Head to think about and the need for system two is no longer a critical mechanism.

Head all to often interrupts Gut and provides a logical reason why there is no danger around the corner. However this does not mean that it will be lost, far from it, this is the system that kicks in when we walk down a dark ally, hear a strange noise in the dead of night, or maybe you are a professional officer and are about to enter a building that you know may contain danger and you feel uneasy. Understanding how these two systems interact with each other is another key in the process to protecting oneself and family. System one uses a quick and simple way of producing thoughts, which we usually refer to as instincts, the process is straightforward and super fast.

Knowledge obtained by Head can transfer to Gut, a novice martial artist learning to strike and kick or a policeman learning to handcuff or draw and shoot, first finds the moves cumbersome and slow, having to continually practice the moves, paying attention to each step in the process, secure one arm with my left hand, reach and find my cuffs with my right, flip them open, snap one side onto the wrist. Continued training and practice, for extended periods of time wires the mental and physical process into the brain, you then come to a point where conscious thought is no longer necessary, you are capable of flowing through the process with speed and accuracy, the process has been internalised, or to put it another way, it has become spontaneous. Interestingly, if at this stage, we were to apply volitional attention to the process, the now fast and spontaneous process would be interrupted and slowed, creating a possible choke point in the learned behaviour.

So system one "Gut" is intuitive, quick and emotional. Gut decides instantly while Head thinks about it for a while, and then finally after life changing seconds have ticked by makes a decision. Gut uses inbuilt settings that are simple rules of thumb, these are hard wired neural pathways that fire when certain stimuli are presented, which natural selection hard wired into our subconscious innate brains a millennium ago, this system does not allow Gut to adjust in any way; it does not give us time to think! These rules of thumb are known as heuristics and biases, they are the brain's way of processing stimuli at lightning speed, insuring that Head does not get involved, putting at risk the survival of the individual.

Similarity: Appearance Equals Reality.

In an effort to understand basic incoming stimuli from our environment, humans use the rule of similarity. This involves grouping together stimuli. Our subconscious brain automatically follows these procedures and in doing so enable it to make very speedy decisions that have aided our survival. The tendency is for humans to group together stimuli that seem similar to each other.

If it looks like a tiger it is a tiger. Like causes like and the brain perceives this, if we see a person being sick just after eating a particular food, we ourselves will not want to eat the same food, this has an evolutionary benefit, as it would have protected the individual from consuming the same food and suffering the same fate. Experiments have shown that if we create a negative thought and feeling this will transfer to our conscious mind and become prominent, over the fact that we know it not to be true. For example take a glass of clear drinking water, apply a label to it that says 'contaminated with radiation', and feel the effects that this will have on you thinking about drinking the water. This rule of like causes like, can be seen when we observe an individual that has in their past decided to ink their body with 'tattoos' for example, or someone that has worked out and is big and muscular.

We link these individuals with bad behaviour or crime and in doing so our minds automatically create a thought of, stay away from them, don't talk to them, there is a threat there somewhere, even though in our mind, if we take the time to think about it, there is no real danger. These linked thoughts of similarity are ancient wiring processes within the brain that are automatically transferred to the mind and brought into conscious thought. In the time of our Stone Age ancestors this process would have worked perfectly, there is a tiger looking for food; we had better be on our way before we are the food!

Framing; It's my Choice

Framing is an example of cognitive bias; humans react differently when having to choose, depending on the way facts are presented to us. The choice will change depending on whether information is presented as a loss or as a gain. As one grows older, the effect of framing becomes more dominant than it is in young adults and children, this could be due to experience playing a role in the thought process. People avoid risk when presented with a positive frame and seek risk when presented with a negative frame. You are looking to purchase a house in a neighbourhood that your unfamiliar with and decide to look at the crime statistics, you are more likely to buy if they are reported as, crime rates are 95% down, rather than crime rate is at 5%.

Anchoring & Adjustment; So Come Up With a Random Figure.

Anchoring and adjustment is another example of human cognitive bias that involves a tendency to rely too heavily on the first piece of information offered when making decisions, this first piece of information is called the anchor. When making a decision, anchoring occurs when individuals use the first initial piece of information to make subsequent judgment on. Once an anchor has been encoded into short term memory, further judgments are made by adjusting either up or down from the anchor, there is then a bias toward interpreting other information around the anchor, it is used for quick, intuitive decisions. Unfortunately, it can result with inaccurate judgments and estimations ranging from taking risk to negotiating the price you are expecting to pay for a second hand car.

Kahneman and Tversky first coined the term "anchoring and adjusting" in a seminal 1974 paper entitled Judgment Under Uncertainty: Heuristics and Biases. They described a consistent two-stage process by which people estimated values. First, an individual establishes a reference point, based on an initial value. This starting point is known as the anchor. Cited by Invisible Law (2013). We then make judgments under uncertainty, based on our anchor point.

Status Quo

Status quo, a bias towards maintaining the current situation. By using a set point, a baseline, any alteration from this point results in a perceived negative loss, this perception often affects the decision making process. How many times have you heard the saying "if it isn't broke, don't fix it"? Taking this a step further, we can relate this to people's decision making, when clearly there is a problem with the path taken, however the status quo has a bias against any change, even if a logical thought process would lead to a different decision.

Sunk Costs

Sunk costs once made are irrecoverable, they represent something that you have already spent and that you will not recover, regardless of future outcomes. This could easily be represented in time and effort, for example you spend many years training in a particular manner, one which you subsequently discover has been maladaptive. Remember the FBI agents learning to draw shoot twice and then re-holster? This was later shown to be potentially life threatening.

Or a long-term relationship that you know is going nowhere. Having to abandon long term attachments or training is of course, easier said than done. There are psychological blocks in place that will not let you simply discard what you have invested so much time in. It is therefore very important that this rule of thumb and the processes involved are understood and taken into consideration when developing training methods, which you intend to rely on for self-protection.

This bias is a hard one to recognise, as all to often we continue to make bad decisions, due to an irrational attachment to the cost that has been invested. There are some very common examples, you take a wrong turning, that you know is the wrong one, but continue for some time, just in case. You keep that investment as it was once worth a lot more than it is now. You continue teaching a self-defence technique or move even though you know there is a more effective one, believing the effort and resources that you have already expended on one path, will be the same amount necessary to alter your path and take a new direction, having to expend the same amount of effort and resources. Never de-value what has gone before, as this past experience has all helped create knowledge, that allows you to understand and acknowledge where you are now, all easier said than done.

Confirmation

Confirmation bias is a tendency for people to favour information that meets with their preconceptions regardless of the evidence or logic that refute the original ideas or beliefs. This in turn creates a tendency for the mind to access memories of knowledge that support your view, rather than ones that could expose a flaw in the knowledge.

This type of bias is more prominent when either emotional or well-established beliefs are involved. If you only look for supporting evidence of your own thoughts and knowledge, you could miss a fatal flaw or a problem that helps solve an issue. Having a confirmation bias towards recalling information can help to explain attitude polarization, when there is a difference of opinion between two parties.

Even though both parties have the same evidence, for example looking at the same scene, each recalls information that confirms their viewpoint, and are unable or unwilling to be objective. After evidence has been presented that clearly explain the answer, a perseverance to still believe persists, this is known as belief perseverance and is seen as helping to strengthen beliefs even in the face of evidence to the contrary.

Ultimately this type of bias can result in decisions that are incorrect or not in a party's best interest, affecting individuals, groups, political parties, military or any social group. Imagine for a moment analysing three different methods of punching using the rear hand, a boxer's cross, a traditional karate reverse and a street fighter's rear punch. Which one of these three is more effective? The answer is at this point irrelevant; it's the process that you used to get your answer that lends itself to confirmation bias.

Cognitive Over Confidence

Another cognitive bias is the over confidence effect; this effect is something that most of us can relate to, especially if you have a tendency for introspection or self-analysation. This is an individual's subjective confidence in his or her own ability, knowledge or judgment. Their confidence is high, even though they later find themselves to have failed or be incorrect. This effect is usually heightened after successive positive results; this then leads to greater confidence that you are right.

Confidence is a necessary trait when it comes to any involvement with violence and therefore should not be seen as a negative one, what has to be kept in check is over confidence. As this is an important issue it will be to our advantage to explore and understand confidence, we need to have a good understanding of this to enable us to: 1, instill this in the people that are relying on you for teaching them life saving techniques; and 2, become mindful of the over confidence effect "a dictionary definition of confidence states that it is "1. firm trust (have confidence in your ability). 2a. feeling of reliance or certainty. 2b. a sense of self-reliance; boldness. 3a. something told confidentially. 3b. the telling of private matters with mutual trust." (Hawkins & Allen, Eds, The Oxford Encyclopedic English Dictionary, 1991, p. 306).

Of these definitions, 2a seems to get to the heart of the matter, as a feeling of certainty can explain definition 1, because trust implies a high degree of certainty in something or someone. Definition 2a also covers definition 2b, as having certainty about one's abilities and attributes. Thus it seems that confidence is essentially a feeling of certainty, the strength of this feeling being the level of confidence, cited by Pulford (1996). What is evident is that this is a feeling within oneself, which produces an air of confidence. It is not surprising that the ego may have a role to play here, intervening on a day-to-day basis to help promote the confidence that we use in our social interactions.

Confidence in your own position can change on a moment to moment basis, never giving an individual the time to consider this evolving confidence, continuing even in the face of logic that refutes your position, now we begin to move into the region of over confidence. I remember being witness one day to a violent confrontation, with two individuals being aggressive to a single guy, together their confidence was high and soon turned from a highly confident position to an over confident position, believing that they had the upper hand. From this position they then decided to escalate their assault to vandalism of the property owned by the individual they were verbally confronting, this was enough to ramp up the response from the single guy, shattering their confidence completely, so much so that they decided to exit the area very quickly.

The over confidence trait exposes the difference between our feelings of complete certainty, which we openly display or share with others, and the cold truth that we are all too often wrong, we continue to defend this position vehemently, even in the face of cold logical truth. The over confidence trait is often responsible for some of the worst parts of human history, playing a part in events that can range from global conflict and war to petty arguments between couples holding their own ground.

However it is not all negative as far as this trait is concerned, having more confidence in your own abilities or your cause can rally others behind you, or send signals to your aggressor that you firmly believe that you will emerge the victor, instilling in others a certain foreboding when confronting you, thus providing an advantage over others. Over confidence would have been naturally selected over and above another trait, leading to those with this effect surviving longer to reproduce and continue to spread this trait in their offspring. Pulford (1996) explains over confidence and in doing so explains that "the most consistent finding is that people are not well calibrated, they tend to have too strong a belief in the correctness of their judgments, that is, they are too confident. When people are over confident they believe that they know more than they in fact do know, or believe their accuracy to be higher than it in fact is". This helps to explain why some lack the ability to see things from another perspective, to a degree they have tunnel vision, again this is not necessarily a bad trait when it comes to survival.

Risk Aversion

Risk aversion relates to a bias for humans to prefer certainty to uncertainty. Individuals that are risk averse have behaviour patterns that are consistent; they tend to stick with their familiar routines rather than venturing into something new. There have been cases where those who are risk averse would create the same routine every day, get up at a certain time, eat the same food every day, drive the same route to work at the same time every day, it could be seen as obsessive behaviour. Risk aversion can have a significant effect on phobias, heightening an already excessive thought process; this would make it very difficult to expose oneself to any situation that would involve a degree of unpredictable responses. When thought processes are over tuned, they tend to be calibrated too far in one direction, restricting normal balanced behaviour, making it difficult for an individual to function. Risk aversion is quite often associated with economics or finance and our reluctance to take risk with money, this gives us an ideal arena to focus our attention on rather than the natural circumstances that offer a risk. Should I walk through that long grass, where I can't see what lies beneath? Is that spider or snake venomous? That food looks old and could make me sick? All carry a degree of risk and even if you reside in a country that does not have venomous spiders or snakes the risk will still present itself.

Adolescence is quite often a period of life when individuals are more prone to heightened risk-taking. In the UK insurance companies consider young male car drivers as having a 100% chance that they will be involved in a road traffic accident, due to the excessive risk that they know will be taken. Adolescents are known for impulsivity, emotional volatility, and risky behaviours. As a person ages, the risk aversion drops, less chances are taken and a certain understanding of one's own mortality begins to appear.

The bias towards taking risk makes sense when applied to survival rather than the markets and a return on your investment. Taking undue risk while living in an environment that had no barriers to being eaten by the predators around at the time, or wandering alone into the wilderness, would certainly have been maladaptive as far as survival was concerned.

Take this risk and project it forward to today's environment where we live in safe, controlled societies and we can come to understand why we have now changed the focus of this risk adverse tendency. Those who work as professionals having to expose themselves to high degrees of risk, for aggressive conflict, with the possibility of danger to oneself, need to understand why they often feel the way they do when taking risk, this understanding will again help to control and deal with the thoughts that the mind automatically creates.

Selective Perception

Perception is the process of attending to stimulus that is received by the mind, interpreting the information and bringing it to the select attention of Head, this is how and when we become aware of the sensory information we experience every second of every day. Selective perception enables the Head to interpret this sensory information in a way that supports our existing knowledge, we already have an awareness of this knowledge and the mind goes looking for information that it has already become aware of; in doing so it supports our existing beliefs.

Imagine if the brain was not capable of retaining past stimulus input? Every single day would be like experiencing a new sunrise, information would never be stored and accessed quickly, a professional police or military person would continually be having to create reference points as to the understanding of incoming perception, we would soon become overwhelmed. The brain is automatically filtering sensory stimulus and only allowing information that it considers necessary to the functional running of the body and mind to come to our conscious awareness.

Over the years, we build understanding about events and information and therefore when selective perception occurs it does not have to re-input and deal with already encoded stimulus, it supports our existing beliefs and knowledge that we take for granted. It's like those old sayings that we hear but do not attribute to anything real "you only see what you want to see" or "none so blind as those who will not see" or " you can't see the wood for the trees" If information does not support existing ideas then it is quickly filtered out, we are the sum of our experiences and any new information that contradicts this is pushed aside.

This process happens on a non-cognitive level; we do not think about it, the experiences that build our knowledge happen gradually over our lifetime. From a martial artist point of view this can mean the difference between sticking with old technique or exploring new and more effective ones, it will depend on your past exposure and in the majority of cases people do not research a particular style prior to taking it up, it's a matter of luck which ones come to your attention.

Once you begin to train, you start the process of building knowledge and belief, your instructor is unlikely to use words that do not support his teachings, they will be constantly supporting the information in ways that you, as a student, will construct as truths. When you are then exposed to information that is potentially beneficial but contradicts existing information, selective perception kicks in and that's the reason that to coin a phrase, that many have a "closed mind". They have already encoded what they perceive as effective, they are limited in their ability to see what is right in front of their eyes and now find it hard to re-wire this thought process.

Counteracting this natural tendency takes a herculean amount of will power, to first become aware of this process in the first place, then to understand your own personal triggers that set off thoughts or feelings that are already deeply encoded within the mind. As you become self aware of this process, you will, over time, be able to create links to ensuring that you recheck new information with already coded material, allowing you to be more open minded and ultimately more knowledgeable and effective in your actions.

Closure

Closure refers to the mind's tendency to see complete figures or forms even if a picture is incomplete It can also be used to help with pattern recognition, partially hidden by other objects, or if part of the information needed to make a complete picture in our mind is missing. For example, if part of a shape's border is missing people still tend to see the shape as completely enclosed by the border and ignore the gaps. This reaction stems from our mind's natural tendency to recognize patterns that are familiar to us and thus fill in any information that may be missing.

Closure is also thought to have evolved from evolutional survival instincts in that if one were to partially see a predator their mind would automatically complete the picture, filling in the gaps and transmitting this to our conscious mind, it would be system one at work telling us that it was time to vacate the area. This can also be linked to the recognition of certain patterns of movement. For example, you are involved in a physical encounter and the attacker is just about to throw a punch at your head, what is the path of action? Is it straight or elliptical? If the mind has already seen hundreds and hundreds of punches coming towards the head then it will of assigned these patterns to system one, which will be discussed in more detail later.

Common Fate

When objects are observed moving in the same direction at the same speed, perception associates the movement as part of the same stimulus. Birds may be identified from their background as one flock, due to the fact that they are all moving in the same direction and at the same speed, even though you cannot identify each and every bird, they may only appear as a small speck. The moving 'specks' appear to be part of a single unit moving in unison. This enables individuals to identify moving objects even when other details are obscured. What can we learn from this? It's very likely that this ability stemmed from our distant ancestors' need to identify and distinguish a camouflaged predator from its background. If we again bring this ability forward a few thousand years and apply it to today's environment, it's evident that unless you live in a part of the World where wild animals still roam, you will not have too many occasions to use this tendency. Of course, the inner cities are often referred to as concrete jungles and getting involved with a mass of people, some of which have you in their sights, may well evoke this tendency to assume that the mass is moving towards you for a reason.

Instincts and Your Intuitive System One.

As we can see from the above, the brain uses a great many varied methods to interpret incoming stimuli and make, in the case of Gut, split second decisions that humans then act upon. If Head gets involved then who knows where we will end up! Instinctive behaviour can also be described as a fixed action pattern, in which a very short to medium length sequence of actions occur without variation, they are carried out in response to a defined stimulus.

Behaviour that is instinctive is acted upon without being based on prior experience; this is further defined by the fact that we have never been exposed to any learning of a habitual act. Instincts can be defined as 'reproductive instincts' or 'instincts of self preservation' A prime example of an instinctive innate behaviour of self preservation would be a new born baby instinctively suckling its mother nipple looking for milk. It's important that we make a distinction here between instincts and reflexes, startle reflexes I will discuss in more detail later.

However, just to be clear here a reflex is a responsive body action to a specific stimulus, i.e. pupils contracting to light. The light changing stimulus causes a reflex where the information does not require brain activity to alter the pupil size, instead the stimulus travels to the spinal cord as a message, which is then transmitted back through the body, tracing a path called the reflex arc. Whereas instincts are patterns of behaviour that are hard wired from birth and cannot be overcome by any volitional effort on the part of the individual. According to Wikipedia (2013) "reflexes are similar to fixed action patterns (FAP) in that most reflexes meet the criteria of a FAP. However, a fixed action pattern can be processed in the brain as well; a male stickleback's instinctive aggression towards anything red during his mating season is such an example.

Examples of instinctive behaviours in humans include many of the primitive reflexes, such as rooting and suckling, behaviours which are present in mammals". In the previous chapter ' To Think What Has To Be Thought', I discussed in some detail volitional attention, it is important here to note that although a fixed action pattern is by definition fixed, it can, by volitional effort, be recognised and adjusted.

Schwartz, and Begley, (2002), in their book, The Mind and The Brain explored the mindful application of volitional attention when dealing with Tourettes sufferers and their 'ticks' which are fixed action patterns. They discuss the power of the mind to shape the brain and in doing so affect the behaviour of the ticks. What they taught patients was a four step process to becoming aware and then modifying fixed behaviour, so that at the point of occurrence they become consciously aware, recognising the point of the behaviour's activation.

This distinctive human ability to take control volitionally of our attention sets us apart from animals, who without a sufficiently strong volitional capacity, are not able to disengage from their fixed action patterns, once they have been activated. In humans, this is distinctive control of system two over system one. According to Lorenz (1963) the human organism has four big instinctive drives that he calls the 'big four' they are; feeding, reproduction, flight and aggression. Any one of these drives also involves multiple small instincts that are designed to support the main drive. In Chapter 5 on violence, I explore the instinct of aggression in more detail.

The dual processing systems of one and two, have been the subject of a great deal of research and there are views putting forward a hypothesis that system 2, Head, is uniquely human. In fact, some go as far to say that Head is what sets humans apart from the rest of the animal kingdom. It is Head that provides humans with the ability to make conscious decisions, to look into the future and the past, use language to tell stories and transfer knowledge via this medium, and create hypothetical thoughts. Another way of looking at this would involve the use of logic and evidence, Head listens to what Gut has to say and then looks for evidence to support the feelings, logic tells us that there is no past evidence to support the feelings and therefore overrides gut, logic right? If it is the case that system 2, Head, is a fairly recent evolutional adaptation, then we are assigning our early ancestors to being nothing but robotic organisms with no volition.

What evidence is there that these two systems of processing are also found in other animals? In a paper written by Evans (2010) he quotes a recent survey conducted into the distinct cognitive systems in a wide range of animal species, he cites the biologist Toates, when he claims that "there is a widespread division between cognition that is stimulus bound on the one hand and involving higher order control on the other. Stimulus bound cognition includes conditioning and the application of instinctively programmed behaviour of a fixed nature. However, he shows that higher order cognition is also present in many species and suggests that this developed into consciousness in humans" A question here would be is system 2 an evolutional adaptation, selected by natural selection, to ensure dominance and survival?

What does Evans (2002) mean when he talks about instinctively programmed behaviour of a fixed nature? System one is responsible for your intuitive thoughts, in the chapter on 'To Think What Has To Be Thought'; I talk a great deal about attention. Attention is a cognitive process, which can be effected by either Head or Gut, we are capable of switching between paying attention to either Gut or Head, and it enables us to make decisions and judgments. Remember, if Gut is first, as it usually is as it operates at speed, then it will bring to your attention pre programmed thought responses, Head then wakes up and jolts your attention to logic. If logic is applied to this process then Gut is the one that we should listen to more often than not, especially if we are involved in an occupation that regularly exposes us to potentially threatening situations and as a martial artist it should be no different.

Teaching students to understand how your brain and mind works also gives them the tools of attending to incoming stimuli that creates awareness. System one, Gut, is automatic, it works with associated memory, it never sleeps and is responsible for quietly controlling and monitoring the incoming stimuli at every moment of every day and to help it do this efficiently it uses the heuristics and bias processes that I discussed earlier. In other words, Gut controls our thoughts and behaviour using a presetting hard-wired into our brains, that is until we undertake a volitional shift in attention and override Gut.

Exploring instinctively programmed behaviour will require us to pay attention to the first thoughts that appear in our conscious minds, not the ones that are introduced by volitional attention. Humans inherit the ability to perceive certain occurrences in our environment without having to have Head get involved.

We recognise behaviour, postures and auditory tones, we react to certain stimuli with a startle reflexive action, we fear situations and other animals instinctively, spiders or snakes for example, or coming close to the edge of a high cliff. System one also learns through repetitive experience creating associated memories throughout our lifetime, using episodic memory to heighten feelings and learning outcomes. Once these memories that are automatically associated with places, events or experiences have been encoded into our brains, Gut accesses them automatically without the requirement for volitional attention to get involved.

In his book "Thinking Fast and Slow", Kahneman, D. (2011) explores system one and two in great detail, introducing readers to the processes that occur within the brain, asking you to think about ideas or thoughts and then attributing them to either Gut or Head, for example;

- Look up and out to a distant object.

- Pay attention and turn to see the origination of a sound

- Complete the phrase, the sky is?

- Make an angry face.

- Detect hostility in a voice.

- Answer divide one hundred by two.

- Read words on large billboards.

- Drive a car on an empty road.

All of these mental processes require no attention to the act of carrying out the exercise, your mind works on automatic. This is Gut at work in everyday activities that we do not have to work at; it's linked very closely to intuition, and happens at speed and with no cognitive volitional attention. Kahneman (2011) also puts forward the idea that "other mental activities become fast and automatic through prolonged practice", we learn them as we grow and experience the world.

Here we have a very important point to consider, is it indeed the case that during everyday activities we instinctively use system 1 'Gut' to speed up the intake of stimuli aiding decisions and judgments'? Kahneman (2011) also states that "system one learns associations between ideas, i.e capital of France, providing the association to Paris, it also learns skills such as reading and understanding nuances of social situations, skills such as finding strong chess moves are acquired only by specialised experts, others are wildly shared" System one is capable of learning in this manner, storing information into memory, which it then accesses without any volitional attention.

The list given earlier is evidence that some of the ways in which we process and answer questions do not require volitional attention, they occur sub-consciously. You are unable to not pay volitional attention to stimuli, it's a little like our in-built startle reflex system, once hard wired, it cannot be altered. Try the exercise with your friends, just insert within your conversation the line 'London has a larger population than the capital of France' they or you will not be able to avoid associating the capital of France with Paris!

Up to now I have concentrated on the internal thought processes of our brain, however, the same is also true for the way system one controls our actions, and just like mental activities that become fast, automatic and hard wired with prolonged practice, the same is also true for our physical abilities and skills.

Over our lifetime we learn to walk, breath, drink, swim or carry out any number of activities, which are eventually controlled by system one. The physical activities that are programmed into our innate action are termed as Neuromuscular Programming (NMP), a subject that will be covered in more detail later. When humans first learn some activities or skills it takes volitional effort. Some are hard wired like breathing, breathing is controlled by system one, yet system two has the ability to jump in and take control of the action, in fact what can happen is that an individual can be in automatic mode, breathing naturally, and then make a volitional effort to pay attention to their breathing and then change their attention to another stimulus, allowing system one to retake control. This continuous, seemingly seamless, transition occurs consistently throughout our everyday lives.

Attention itself is shared between both systems. There are also skills that initially are not hard-wired or automatic and, like the specialized expert in chess, these skills have to be practiced over prolonged periods of time. Let's take an example of a skill that requires instinctive reactions that involve speed and complex motor skills, the skill is not really the important thing; it's the process that is important. Any new skill could be used to illustrate my point here, so I decided to choose the drawing of a gun and controlling a suspect, as this behaviour has several psychological and physical actions that will repeat themselves throughout this book.

Imagine that you are a rooky entering the training centre for the first time, eagerly awaiting your first lesson on how to draw your sidearm and control a suspect. Up until this point you have never used a handgun, let alone pointed it at anybody. The instructor goes through all the required safety briefing, explains what he wants you to do, to start with lets keep it simple, reach down unclip the safety strap, draw your weapon, aim and take up a standing position ready to issue commands, easy right! Off you go to practice the skills.

The first thing that you will recognise is that you have to think about each step in the process, you have to pay attention to the safety strap, drawing the weapon slowly, smoothly, holding it correctly, taking up a position and aiming the gun. Every part in this process requires your full attention; you think to yourself that you will never be able to do this at speed with the adrenal dump happening as well.

What is different in the above, other than the physical action from learning what the capital of France is for the first time? You have to pay attention and focus on remembering the capital's name, just like you have to remember each step in the drawing of your weapon, this time with a corresponding physical skill as well. It's a given fact that if you continue with prolonged practice, over time you will build the ability to perform these moves without thought, but to begin with you have to pay attention, focus, concentrate, any of these terms are relevant, you will then over time consign the attention and ability to system one.

At first your moves will be slow and deliberate, over time they will become faster and eventually, as a specialized individual, you will be able to draw at speed, allowing system one to take control of the skill. If, for any reason system two gets involved with thinking about what you have to do, then the whole process will go into slow down mode, this is no different from being asked a question that you know the answer to, yet for some strange reason, Head got involved with thinking about the answer and 'for the life of me I can't remember! It will come to me!

Don't think about it'! Sound familiar? This is one of those times where system two overrides system one, this leads to errors in decision making, judgments and of course the carrying out of physical skills that were originally assigned to the realm of spontaneous movement. Then there are those all too critical times that we fail to listen to system one, when we should use system one to override system two, those times when your Gut tells you that something is not right, the hairs on your neck stand up, you go all goose pimply, just like my ranger friend when he never listened to his instincts and suffered the consequences of not overriding Head with Gut.

All of these processes involved in learning a skill or thinking critically require a degree of attention and effort, it will not just happen on its own. The more demanding the task the more effort has to be attended to, if the effort involves system two getting involved then be prepared for a slow down in your normal speed of operations. For an individual involved in high stress situations on a regular basis, knowing what can impact upon performance is critical, for the individual that trains for a possible encounter at some time in their life, never knowing if such an event will ever occur then having the ability to access your instincts will be a distinct advantage, it will aid your performance when it really matters, using system one to create super fast thoughts and reactions.

I discuss "performance under stress" in Chapter 9, and now want to explore the effects of conflict when systems one or two are not working congruently, as they also have a significant effect on performance. There are two types of degradation in behaviour performance that effect an individual's decision-making and physical abilities; according to Kahneman (2011) they are connotative load and ego depletion. Experiments by psychologist Roy Baumeister provided evidence that system two controls thoughts and behaviour, and volitional effort to control any connotative thoughts, emotional responses or physical actions, requires a large amount of mental energy.

The depletion of mental energy will soon affect any individual, slowing down future attempts at skills that can usually, without mental distractions, be carried out with speed and efficiency, this is known as ego depletion. The tasks that will effect performance are varied, however the common denominator between each task or skill is conflict, leading to a potential for the loss of motivation for up and coming tasks.

Cognitive load requires the continued attention to a mental task, keeping a thought in your mind, system two is at this time engaged in activity of mindfulness. As a result of these experiments Bumeister also discovered that the central nervous system, while maintaining high levels of mental activity, uses larger amounts of glucose. The amount of mental activity that you undertake is synonymous with a drop in blood sugar levels, the harder and more prolonged the activity the lower the levels fall. This has a significant implication on an individual's ability to perform at optimum levels, could it be the case that an intake of glucose would counteract the loss of mental and physical abilities?

Kahneman cites experiments published in the Proceedings of the National Academy of Sciences, they detailed the performance of eight judges in decision making when they were tasked with awarding or denying parole. The judges spent the whole day reviewing cases and making decisions as to whether to grant or deny parole applications, their default position is to deny if there is any doubt. They take a short amount of time on each case an average of 6 minutes and they are presented in no particular order. 35% of the cases are approved, the exact time of each decision is recorded along with the time that each judge takes a food break, one in the morning, one for lunch and one in the afternoon.

The researchers plotted the number of approved applications against the time when the food break was taken, after each meal the proportion of approvals spike 65%, of all approvals are taken just after the food intake. During the period of time when the break is over, until the next food break the approval rate drops steadily throughout this time and reaches zero in the immediate time before the next break is scheduled. This research indicates that the judges began to make the easier default decision as they became more tired and fatigued.

This has a direct implication to high-level events when physical exertion and decision making are tied into situations where high stress is also present. For the professional, understanding the core reasons behind fatigue, poor decision making and tiredness is again critical to optimum performance. It's not rocket science to know that a lack of food will effect performance and decision making, most of us have experienced a lack of energy or being unable to think straight at some time in our lives due to a lack of food, this is about detailed understanding of what is happening to us when under stressful situations and being able to cope with its effects.

Within this chapter I have explored the two systems that we use to navigate our environment, making decisions and judgments as we go. Understanding how our brain works allows us, as individuals, to become aware of the processes involved and once we are able to do this, mindfully pay attention to the process and modify it in some manner. Before I move on, the last innate mechanism that I want to explore is that of Priming and although we may not like to admit it, we are all subject to this very subtle mental process, that can have a significant effect on our behaviour and our psyche.

Priming

Priming an individual creates an increased sub-conscious awareness to certain stimuli, creating a link to prior experiences and knowledge. Priming is not memory retrieval; this would require active access to past memories and experiences. The brain recalls explicit memory that focuses on words and objects, priming was originally thought to revolve around words and visual patterns, however later research found that priming also had an impact on social behaviour, it can help guide or channel behaviour within a given situation, "it has a causal effect on goals to achieve high performance, to cooperate with an opponent, or to be fair minded and egalitarian" cited by Bargh (2006). In the context of aggressive hostile behaviour, priming can create a stereotype, influencing the observer's interpretation of what they think they see is happening, being able to prime individuals before they undertake a confrontation or a hostage negotiation for example, could cause a significant shift in behaviour.

A prime helps to activate particular representations or associations in memory just prior to carrying out an action or task. For example, a person who sees the words "calm, control" prior to a confrontation will be more inclined to be empathetic to someone with a problem, their behaviour and mannerisms will reflect a person who is calm and in control. This happens because the words 'calm and control' are closely associated to experiences that support this type of behaviour. Priming can refer to a technique in psychology used to train a person's memory in both positive and negative ways, leading us to the conclusion that priming is the sole domain of system one.

What the brain is doing here is associative activation, the thoughts that you have created in your mind using system two, are sub-consciously effecting the brain. System one is behind the scene and creating all sorts of interpretations for the words, and you, the individual, have no volitional control over this process. In your mind's eye you formulate an idea, thinking about aggression will automatically trigger this process, and it is a very coherent and congruent process. The word will evoke memories, which then leads to the generation of emotions, this then leads to real physiological responses by the body, triggering reactions that then support the initial thought, your heart rate increases, you tense up, your facial expression change, causing micro expressions, sub-conscious facial movements that are again completely subconscious and are an indication of the silent workings of the brain.

System one is in full flow and working at lightning speed, creating patterns of thoughts that reinforce each other. In the context of real life situations, this brings forth both visual and auditory stimuli and ensures that the priming effect will be more powerful, the emotional and physical reactions even more intense than just sitting in a quiet room and thinking about or reading the word 'aggression'. Kahneman (2011) adopts an expansive view of what an idea is stating "it can be concrete or abstract and it can be expressed in many ways, as a verb, as a noun, as an adjective or as a clenched fist". Any idea is like a game of chess, once one move has been made it will impact on a string of different ideas and possible moves.

The important understanding here is that the mind is neurologically programmed in various different ways, when we use language to influence the mind we call it Neural Linguistic Programming (NLP), when we think the idea it's Neural-logical Programming (NP) and when we create a posture with our body its Neural Muscular Programming (NMP), all three processes are consistently interactive in priming, they are transferred from the workings of our brain to our conscious mind via system one, automatically and at speed. The implications and uses for the above programming, their effect on the mind and the body, will be explored specifically later, getting to grip with the term NMP will help increase our understanding and our ability to create efficient and effective body mechanics, especially when it comes to conflict, aggression and the manipulation of our neural pathways.

3 MEASURING LIFE BY THE NUMBER OF BREATHS WE TAKE

The poet Maya Angelou is credited with saying, "Life is not measured by how many breaths we take, but by the moments that take our breath away."

Meditation requires an individual to train the mind, there are several terms that can be related to meditation; concentration, focus and attention, all involve an amount of self discipline over the thoughts that intrude the mind every micro second of everyday. A term that has become popular recently is "mindfulness,' this is a particular state of consciousness and can be brought to bare on thought processes of the mind, a way of examining the modern mind and the ways in which we interact with ourselves, those around us and our environment. It provides us with a method of examination and analysis of our experiences. The terms meditation and mindfulness could easily be interchangeable, both terms refer to a variety of methods in which the individual pays attention to various thoughts that are designed to promote relaxation and a certain power of the mind over the body, it is a state of consciousness.

Back in the 1800s, one of the fathers of psychology wrote, "Millions of items of the outward order are present to my senses which never properly enter into my experience. Why? Because they have no interest for me. My experience is what I agree to attend to. Only those items which I notice shape my mind, without selective interest, experience is an utter chaos" James, W. (1890). Meditation therefore requires a certain amount of volitional attention to specific incoming stimuli, to the exclusion of unwanted thoughts or body states, it is a general term that encompasses methods of focused concentration on a particular object, body state or internal thought, that can be achieved through disciplines like the martial arts, yoga or psycho analysation. a

Meditation is usually undertaken while at rest, sitting or standing without movement. Mindfulness is more of an open monitoring process, where your intention is to be mindful of your body and mental states; it can also be accessed while going about your daily lives, walking, gardening or during your occupation. The last one is of particular importance if your occupation involves exposure to high stress/emotional situations.

The practical method of meditation/mindfulness can be achieved in a variety of general ways that include visualisation, tactical performance imagery, positive self-talk, relaxation, de-stress and concentration. There are also methods that are linked to religion, such as prayer or chanting, it can be practiced while in movement, including walking, yoga or a martial art like Tai Chi. The majority of people today associate meditation with the sitting posture of the half-lotus position, however stationary resting meditation can be practiced in any posture.

If you are not convinced that being mindful and using techniques such as tactical performance imagery to enhance your training and skills is a valid method, this extract from the true life story of a guy that has seen more combat than most may help "I would visualize the whole attack, going through the entire motion but without a weapon in my hand. It's like shadow boxing. You practice a fluid motion, but the mental aspect of it is as important as the physical part. You have to visualize whatever type of knife or weapon you will be carrying in your mind; you have to make your movements with that weapon, knowing how long it is and how heavy it is, whether it's a pocketknife, a machete or a submachine gun. You have to practice planting your feet and bringing all your emotion within your body into delivering the KILLING blow" O'Neal (2013)

One of the key aims is to develop a life force or internal energy that can be directed towards obtaining your goals. This life force is also known by several different terms depending upon the culture that you live in. In the Far East it is called chi or ki, in India Praná. Buddhism is probably the most well know religion that practices meditation, known as bhavana. The practice of meditation has been around as long as humans have had a sense of self and the ability to volitionally pay attention to their thoughts. Volitional attention is mindfulness, as long as an individual can hold a thought in their mind, whatever it is, they are paying volitional attention to themselves, I will come back to this later, this I refer to as Volitional Mindful Attention (VMA) and is a key application when undertaking Volitional Attention Training (VAT). Meditation has a symbiotic link to many religions of the world, helping to create belief, to help develop emotions and traits, happiness, love, compassion, benevolence, forgiveness, the list goes on, it also means different things to different people depending on the context.

Meditation requires self-regulation to become mindful of one's own thoughts, however it's more than just your internal thoughts that are relevant in self-regulation. It also has a significant effect on regulating and adjusting our behaviour, emotions, feelings and thoughts. It enables individuals to control themselves, adapt to external stimuli within our environment and the context in which they are presented as well as achieving goals. Meditation is also used to clear the mind and manage all sorts of health problems, such as high blood pressure, depression, and anxiety. In the context of this book, it's obvious that the benefits of meditation in accomplishing self regulation is a vital skill for any warrior or martial artist, all too often meditation is only seen as a holistic benefit to people's lives and wellbeing, the benefits of mindfulness to the warrior are usually overlooked and seen by many as the realm of the soft arts like Tai Chi. Consider the following, "the ability to concentrate and focus is the most intuitively recognized essential skill for responding effectively in high stress situations" Asken (2010).

The ability to concentrate is enhanced by the skill of meditation, training in mindfulness is the practice of concentration "Combat requires the ability to focus like nothing else" Navy SEAL Machover, cited by Asken (2010). What is evident, is that mindful self-regulation has a significant role to play in performance during high stress situations. Just knowing that mindfulness is a small part in the cog that drives behaviour in emotional situations is not enough, it has to be trained, just like throwing that right cross, snapping on the handcuffs or drawing and firing your weapon. It is one of the least practiced skills in the combat arena, yet it is evident that it is one of the most important of skills, it is the one that keeps your mind engaged and focused, it's the oil the enables the mind's neural connections to fire and remain in control when everything else is shutting down. Stress and emotional fear are constant companions when engaging in potentially life threatening situations, the best tool to cope with these is training and self-regulation.

Being able to monitor and adjust your own body state while engaged in high stress situations will provide an individual with an advantage over those that do not possess this skill, one really important point here is that any benefit that could be gained from mindful self regulation will be lost if the body has to cope with any counteracting substance, medication, alcohol, drugs or psychological problems as these will all severely degenerate the ability to self regulate oneself.

Mindful meditation, as we have seen with volitional attention will have a biological effect on the brain, this state of mind can, over time, and with the appropriate amount of training, become more of a trait than a state of mind. Sue Smalley and Dian Winston (2010) discuss mindfulness as a state and discuss the fact that this state can be changed through a wide range of experiences. They quote a large amount of research that points to the fact that mindfulness practice actually changes physiological states.

For example, the immune system gets stronger as reflected by the increase in the number of cells fighting infection, and brain function changes, with a measurable difference in the amount of grey matter in particular regions of the brain within long time meditators', compared to those that do not meditate. The overall result of long-term mindful meditation is a healthier individual, with a sense of wellbeing, less stress and anxiety leading to a calmer personality. What we see here is that long term meditation has a real effect on the brain structure and in turn, just like volitional attention can help to rewrite the neurological pathways of the brain, adding and affecting the creation of neural plasticity.

OODA Loop

This term was originally proposed by the United States Air Force Colonel John Boyd for combat operations, the OODA (Observe, Orient, Decide, Act) Loop has since been applied to many different endeavours, these include, business, sports and politics, both on group and individual levels. The individual having to go through the OODA loop when involved in a high stress situation can mean the difference between success or failure, any setting where information needs to be analysed and decisions made relatively quickly involves this process.

The concept can be used to explain the mental processes that are involved in the strategy of war, to defeat an adversary and survive. Applying this process to a soldier that is constantly under pressure to change and adapt to different incoming stimuli requires the loop to be continuously used. What can be learned from the OODA loop as a model for how we should think about near real-time decision making, is that in theory, it is a logical process, albeit a slow one.

"Boyd's model for human decision-making is that we observe that something has happened; place that observation in the context of the other things we know about the situation; decide on a course of action; and then act on it. The process is a loop because we continue to gather new information and make new decisions as the situation unfolds" The TIBCO Blog (2013).

Even though the logic is clear with the OODA loop, the actual decisions that are made will in a great many cases, be built on the brain's processing mechanisms, such as the bias rules discussed in chapter 2, the bias of similarity, priming and anchors. What system initiates the loop, system 1 or system 2? Again logic would answer that it's a volitional choice to decide and then act. Enhancing this process with the use of mindful self-regulation will aid the speed of the decision and if stress should raise it's head, then the ability to breath and calm the mind will aid in helping the OODA process.

Being able to short cut the speed of decisions will enable an individual to interrupt an opponent's thought cycle, therefore providing a clear advantage, just like the experienced chess player, who is always several moves in front and the further in front you are, the bigger the advantage.

To enable an individual to self regulate performance under stress, to be mindful of cognitive processes, body states and emotions requires a thorough understanding of mindful meditation and one of the most commonest ways to become mindful and begin to self regulate is to learn how to breath. There are a multitude of techniques that enable an individual to practice meditation so let's start with one of the most common ones, breath.

Breathing

One of the easiest and quickest ways to learn mindful self-regulation and also provide an anchor to behaviour is the control of your breath, before I look at some practical applications let's look at the science of breath.

The act of breathing is for the most part under the control of the Autonomic Nervous System ("ANS"); the region responsible for the control of breathing is the Medulla within the brain stem. The ANS is responsible for the monitoring and control of our internal environment; autonomic suggests "independent" of the conscious mind. However, we are all capable of volitional control of the system, it's similar to the control exercised over our mind and brain by system one and two discussed earlier, for the majority of the time throughout our lives the function of breathing is completely automatic. There are times, and meditation is one of them, when an individual volitionally takes control of breath.

The central nervous system (CNS) of humans consists of the brain and spinal column and is responsible for all of the input data and communications throughout the body. To communicate to the extremities of the body we have the Peripheral Nervous System (PNS), this in turn is divided into the Somatic Nervous System (SNS) and the Autonomic Nervous System (ANS). The ANS is further divided into two subsystems: the Parasympathetic Nervous System (PSNS) and Sympathetic Nervous System (SYNS), which operate independently or co-operatively depending on the requirements and control of the brain. These two sub systems are dichotic in their control of homeostasis within the body for example; one activates and excites a physiological or body state response and the other inhibits them.

The two systems SYNS and PSNS act in a dichotic manner as well, they are responsible for firing up or slowing down the internal operating systems within the body, the SYNS which is a "quick response mobilising system" fires the body into action and is responsible for the fight or flight response discussed earlier, the PSNS system is a "more slowly activated system", which is responsible for bringing the psychological and body states back to normal operating homeostasis control levels. The SYNS is responsible for the delivery of oxygen and nutrients throughout the body and has been found to sometimes overwhelm the homeostasis of the body by infusing energy, this can cause hyper arousal, that leads to dizziness, fear, and over excitement, it is the PSNS that rains in the SYNS and brings the body back to a state of equilibrium in homeostasis within the body. Breathing is the one constant that can control hyper arousal, which if unchecked in a high stress or combat situation, can again have significant effects on psychological and body states, leading to mental and physical shutdown.

The scientific name for breathing is ventilation and there are two phases to this process, inspiration and expiration. While we are in the cycle of inspiration the diaphragm and our intercostal muscles contract, lowering the diaphragm down, which in turn increases the amount of volume in the chest, our rib cage also increases in size, this has the effect of lowering the air pressure within the lungs, this creates a pressure differential between the inside air pressure and the external pressure and this results in air being drawn into our lungs.

This cycle then returns to a resting state, the diaphragm and the intercostal muscles relax, which forces the air out of our lungs. It's important that we do not confuse the term ventilation with respiration; respiration is the activity that occurs within the lungs when the exchange of oxygen and carbon dioxide occurs between the air and the blood within the lungs. Oxygen is transferred into our red blood cells, to be transported around the body via the arterial blood system, at the same time carbon dioxide is transferred back into the air, which is then disposed of back into the atmosphere. The oxygenated blood travels throughout the body literally feeding our muscles and brain; once the oxygenated blood has been depleted it is returned to the lungs via our veins where the exchange occurs.

This continued cycle delivers oxygen and nutrients to virtually every organ within the body, including the brain and musculoskeletal system; and keeps the body in a state of homeostasis. If the balance is upset and for example; an over abundance of oxygen is consumed by the lungs and transferred into the body system, more energy than is needed is created, this can cause dizziness, fear, anxiety or other forms of hyper arousal and distress, this is when hyperventilation can occur, this is a state that should be avoided at all costs, especially when having to deal with a high emotion state of arousal.

The average human breath cycle is twelve to fifteen times a minute; with each breath we consume nearly two litres of air. When mindful control takes over the cycle of breathing, it alters our respiratory sinus arrhythmia, unfortunately not all the air we intake into our lungs is the best of quality, we have the potential to also breath in pollutants that can affect the efficiency of the whole process, smoking is an obvious example, however there are many more. What is less known is that internal processes can also significantly affect the efficiency of this system, anxiety, stress, panic, excitement, disease or infections; they can all take their toll on the efficiency of our breath. When confronted with a high emotional stress situation our breathing rate can alter dramatically, the heart rate increases, we sweat profusely and quite often high chest breathing occurs and we can go into hyperventilation.

Hyperventilation, is caused by rapid quick breaths that do not allow time for the lungs to perform their function, exchanging oxygen into carbon dioxide, resulting in excessive ventilation of the lungs, instead of a steady slow inhalation of oxygen and exhalation of carbon dioxide, this manner of breathing creates low levels of carbon dioxide in an individual's blood, which in turn leads to the feelings associated with hyperventilation.

It is generally associated with psychological states of emotion, it is also known as 'behavioural breathlessness' and is a result of a stimulus rather than a cause, and can have a significant effect on performance, especially during situations when fear is felt. It therefore needs to be recognised so that an individual experiencing symptoms during encounters will have the knowledge and skill to deal with the state.

If you regularly experience this type of breathing cycle you may have Hyperventilation Syndrome (HS). Hyperventilation syndrome can cause an individual to experience heightened emotions of stress, anxiety, depression, or anger. A specific fear or phobia can also trigger a HS event, which can range from, claustrophobia to encountering a spider in the washbasin. What is important here is the recognition that a reaction to this type of event is completely normal, having a process to deal with the resulting increase in breathing and heart rate is what self-regulated mediation is about.

Slow deep breathing is a process where an individual can take back control of the ANS, specifically the parasympathetic nervous system, which is responsible for returning the body back to a calm homeostatic state. "mindfulness may be a method of enhancing the capacity of the parasympathetic nervous system to bring the body back into a homeostatic state, the mind/body relationship is bi-directional, the mind can influence the body and the body can influence the mind" Smalley and Winston (2010). This clearly provides us with the importance that should be placed on the practice of self-regulated, mindful meditation, breath control is therefore a key skill and one with which every professional or martial artist should be familiar with, if not then you put your performance and potentially the life of your fellow professionals at risk.

Breathing is a dominant process within all humans along with heart beat "your heart beats about 100,000 times in one day and about 35 million times in a year. During an average lifetime, the human heart will beat more than 2.5 billion times" NOVA Online (2013).

Your breathing amounts to " The average respiration rate for a person at rest is about 16 breaths per minute. This means on average, we breathe about 960 breaths an hour 23040 breaths a day and 8,409,600 a year. If a person lives to be 80, then on average they will take 672,768,000 breaths in a lifetime!" Answers (2013).

Both of these processes are so automatic that for the most part we never pay much attention to them, until something goes wrong or we find ourselves in a state of fear or excitement. Due to the complete dominance of breath within our body it is the simplest and most reliable process that we can turn our attention to, after all, if we are not breathing the chances are we are on our way to the afterlife and as long as you have a breath in you, you are alive!

To be able to take our volitional attention and turn it to the control of our breath is no different from training any physical move, we have to take time and train our attention. Remember that any time we take our attention and focus it on an event it costs effort. The aim of this training is again no different from the ultimate aim of combat and martial arts, and that is to make the training spontaneous, not requiring effort to perform, you are in the groove, or zone, so to speak.

While in this state you are completely emerged in the moment and your chosen activity at the time seems to flow without any cognitive effort. In Chapter 1, To Think What Has to Be Thought, I talked about the research done on the brain while cognitive effort was involved in learning a task, then later the brain regions responsible for this learning process had shut down, well mindful meditation is no different.

Sensory acuity

Acuity occurs when individuals train certain senses and behaviour to a degree of expertise, defining what makes an expert in a certain activity is difficult as the parameters for measuring expertise are vague. Time and experience, might be one measure, however I have already discussed the possible errors that can occur in teaching and training behaviours that could be seen as maladaptive to your field of expertise.

In general, sensory acuity requires training in a particular field and can involve all five basic senses within the human body. A chef hones his ability to define taste and can distinguish the difference between many different ingredients; a perfumer has the nose to sniff nuanced fragrances and a superlative sense of smell, providing the individual with the ability to identify scents with precision.

A musician has the ear to create orchestral masterpieces; a blind person the ability to decipher a closely arranged multitude of dots on the surface of paper and interpret them into words, and an artist has the ability to see colours and composition, to create a visual masterpiece. Today, modern scientific understanding of the human body and the 5 basic senses has expanded the number of senses within the body, there is now no longer just the big 5 and depending upon what you read, the new number of senses range from the standard list of 5 senses to 14 and 20 different senses.

A short definition needs to be understood in order to provide us with an understanding of why this number has now been significantly increased. To be able to sense something both within our bodies and in our environment requires a sensor of some description and depending upon its function, will mean it has either one specific job to do or it gathers a multitude of incoming stimuli.

For example, your eyes detect light through two different types of sensors, 'rods' work in low-light and detect light intensity 'cones' require intense light and detect colours, there are three types of cones, one for each of the prime colours. So although sight falls under one category, there are two senses that make up the one and one of those is subdivided into three.

Our skin is the barrier between ourselves and the world around us and as such is one of the main sensors to incoming stimuli and has five different types of nerve endings that are independently sensitive to heat, pain, itch, cold and pressure, they are responsible for providing us with a sense of temperature, pain, touch and itch.

Our sense of smell can bring on a flood of memories that effect our emotions and moods also known as our olfactory system and is part of the brain's limbic system, an area associated with memory and feeling. Smells can evoke strong and vivid memories that are capable of activating the body's reflex system to protect itself, it's just like the wild cat with its nose in the air detecting its prey and any potential danger from smelly humans.

Within your muscles and joints, there are sensors that provide you with awareness information as to where your body parts are within space and time. These sensors also allow control of movement and tension that enables complex locomotion and co-ordination skills, this internal sensor system is discussed in greater detail within the chapter on Neuromuscular Programming.

Having the ability to be mindful of your internal and external states will provide a degree of self-regulation over your body, training particular sensory acuities will also allow for a heightened awareness in certain situations. As a professional, either in the field of security, police or the military, training a heightened sensory acuity that enables faster responses to potentially life threatening situations should be on the list of required skills to perform your job effectively. This method of training will help enhance your ability when exposed to real time encounters. Volitional Mindful Attention is a skill that should be trained alongside any practical skill set, the difference is that you need to pay attention to sensory acuity to help you survive and respond to violent and aggressive encounters and not, as with most meditation practices, relax you to a state of stillness within the mind and your body, although this is not a bad thing, as long as it's done within the correct context, going into a relaxed state may not be ideal when having to deal with an armed aggressor.

Training our attention

There are specific regions of the brain that research has shown to be active during meditation. "Buddhist monks who do compassion meditation have been shown to modulate their Amygdala, along with their Temporoparietal junction and Insula, during their practice. In an FMRI study, more intensive Insula activity was found in expert meditators than in novices.

Increased activity in the Amygdala following compassion-orientated meditation may contribute to social connectedness" Wikipedia (2013) Amygdale. Here we find evidence that science has been able to bridge the gap between mystic meditation by monks and the actual effects that this type of self-regulation has on the brain, let's look at some of the practical methods of meditation.

Methods of Practice, Pranayama.

Certain types of meditation and yoga practices use Pranayama breathing; they advocate the practice of volitional breath control. This type of breathing requires a practitioner to inhale, retain and exhale quickly or slowly. Sovick (2000) considers this type of breathing to be an "intermediary between the mind and body". Previously I identified the word 'prana' and referred to it as the 'life force' or energy that all humans and indeed many would argue, all living organisms have. Breath is responsible for the intake of oxygen, which then via the blood stream disseminates this energy containing substance to all parts of the body, depending on the consumption requirement.

The brain requires approximately 20% of the total energy of the human body which compared to its size is a very large amount. Sovick (2000) later says, "There is a direct connection between the 'prana' or energy of breathing and its effects on energy liberation in the body. Cellular metabolism (reactions in the cell to produce energy) for example, is regulated by oxygen provided during breathing". Yoga practices a slow control over the breathing process in order to generate a greater feeling of energy and relaxation throughout the body, to control the body states, to focus and clear the mind and to become aware of the internal working of the mind and body. "Pranayamic breathing, defined as a manipulation of breath movement, has been shown to contribute to a physiologic response characterized by the presence of decreased oxygen consumption, decreased heart rate, and decreased blood pressure, as well as increased theta wave amplitude in EEG recordings, and increased parasympathetic activity accompanied by the experience of alertness and reinvigoration" Jerath (2006). Jerath also states that pranayama breathing has been shown to positively affect immune function, hypertension, asthma, autonomic nervous system imbalances, and psychological or stress-related disorders. Investigations regarding stress and psychological improvements support evidence that pranayama breathing alters the brain's information processing, making it an intervention that improves a person's psychological profile. This evidence points to a clear process that can be trained, enabling individuals who are exposed to difficult fear producing situations, to control both psychological and body states that could severely impact on performance.

Tactical Breathing

This method of breathing is not unlike any other, its name however "tactical breathing" is synonymous with combat and high stress situations, Asken (2010) talks about tactical breathing as being useful in managing the arousal or stress of a mission, he cites Siddle (1995) ' we would argue that breath control should be a mandatory component of survival stress management", powerful support for the activity of mindful meditation. There is no real big secret here, it's just paying attention to breath, meditating, being aware of your own body and mental state.

One method of tactical breathing is described by Grossman in his book On Combat (2004), this he describes as the 'four count'. Begin by breathing in through your nose to a slow count of 4, which expands your belly like a balloon. Hold for a count of 4, and then slowly exhale through your lips to a count of 4, as your belly collapses like a balloon with its air released. Hold empty for a count of 4 and then repeat the process. Remember that part of this whole process is to create a more focused mindful state, to control any stress or fear that may well be beginning to take hold of your thought process. This is not about taking five minutes to calm yourself and relax, it's about creating an anchor mechanism attached to a thought process that allows you to manage the high emotional situation you find yourself in and do not think for a moment that this can be done 'just like that'! It's going to take some time and effort on your part to train this type of mindful breathing.

It's important that we remember that what we are doing here is taking control of our autonomic nervous system and using this control to self regulate our mind and body states, for the majority of the time our bodies are on auto pilot, the reason for bringing meditation into this subject is due to the fact that you cannot be at your best unless you have control over your self, breathing is your bridge between the somatic and autonomic nervous system, Grossman (2004) puts it well " Tactical breathing is a leash on the puppy. The more you practice the breathing technique, the quicker the effects kick in, as a result of powerful operant and classical conditioning mechanisms" One thing is for sure no longer is meditation relegated to the realms of the Buddhist monks.

Brain States While Being Mindful

The brain never sleeps, or more accurately, we are mentally active, resting or asleep, the brain always has some level of electrical activity. Humans have 5 different states of brain rhythm activity, each occurs at a different frequency. Brain waves can be monitored through the use of electro encephalography (EEG) and depending on whether we are awake, active, awake at rest, deeply asleep or during Rapid Eye Movement REM our brain waves will alter.

Research has shown that during meditation, theta waves were most abundant in the frontal and middle parts of the brain according to Wikipedia (2013) "cortical theta rhythms observed in human scalp EEG are a different phenomenon, with no clear relationship to the hippocampus. In human EEG studies, the term theta refers to frequency components in the 4–7 Hz range, regardless of their source. Cortical theta is observed frequently in young children. In older children and adults, it tends to appear during drowsy, meditative, or sleeping states, but not during the deepest stages of sleep. Several types of brain pathology can give rise to abnormally strong or persistent cortical theta waves"

Beta waves within the brain are present when the brain is cognitively awake, there are three different rhythms, low beta, beta and high beta, with low frequency beta related to a very active and engaged brain with high levels of attention for example; writing a letter or doing a puzzle of some kind. During this state, a heightened state of alertness, logic and critical reasoning also occurs. An overall higher level of beta levels occurs at the same time that we experience stress, anxiety and restlessness. The majority of adults are in a state of beta throughout their normal working day and it is while in this state that stress is produced, counteracting this state will provide a certain level of relaxation and less stress.

Alpha waves occur when you are in deep relaxation usually with the eyes closed and while daydreaming. There has been research into alpha waves that suggest that there are three types of wave, as they have been detected during REM while sleeping. The third type of wave is known as alpha-beta waves. Alpha beta brain waves are present while in a state of relaxed detached awareness and can be achieved through light mindful meditation. During this state of mind you are in the optimal state for programming your mind for positive results in any activity. Alpha heightens your visualisation, memory, focus, learning and your ability to be volitionally self aware, it is said that this state is the gateway to your subconscious mind.

Delta are present when we sleep, particularly when we are in a state of deep sleep, they are the slowest frequency of brain wave activity that we experience. This wave frequency is found during deep dreamless sleep and also while performing transcendental meditation where our awareness is detached from our connotative minds. Delta is the realm of your unconscious mind. It is hypothesized that while in this state, regeneration and healing occurs, during this state we are able to access information stored deep within our minds.

Gamma waves.

These brain wave states are pertinent to the subject of this chapter and have an implication throughout this book; we should therefore take some time to understand the gamma frequency. A gamma wave is a rhythm of neural oscillation with a frequency between 25 and 100 Hz, although a frequency of 40 Hz is the more typical brain rhythm. The gamma wave is particularly important to the topic of meditation.

There have been experiments conducted on Tibetan Buddhist monks that have shown a link between transcendental mental states and the gamma wave frequency. In a recent study Ferrarelli (2013) and colleges conducted research into how sleep provides a unique approach to explore the meditation-related plastic changes in brain function. While the study participants were asleep the team recorded high-density electroencephalographic (hdEEG) recordings in long-term meditators of Buddhist meditation practices, with a mean practice hours between them of 8700 hours and meditation naive individuals. The results indicated that the long-term meditators had increased parietal-occipital EEG gamma power during NREM sleep. This increase was specific for the gamma range (25040 Hz), it was not related to the level of spontaneous arousal during NREM and was positively correlated with the length of lifetime daily meditation practice.

These findings do indicate that meditation practice produces measurable changes in spontaneous brain activity and does suggest that EEG gamma activity during sleep represents a sensitive measure of the long-lasting, plastic effects of meditative training on brain function. Studies were also conducted on meditation experts and novices during resting wakefulness just prior to meditation; the results were the same, an increase in gamma power in the same region of the brain. Ferrarelli's (2013) research returned the findings that support their hypothesis, they found that at the individual subject level there was a significant increase in gamma power, there was more than a 50% separation between long-term meditators and novices. There was also a significant correlation between parietal-occipital NREM gamma power and daily meditation compared to retreat meditation or novices.

Expert Buddhist practitioners with more than 10,000 hours practice in volitional focused attention as well as open monitoring meditation were also compared to a group of novices. The expert practitioners showed self-induced, higher-amplitude, sustained EEG gamma-band oscillations, especially over lateral frontal parietal electrodes, while meditating as well as in the resting state immediately preceding and following meditation. Notably, a link between higher gamma-band activity and stronger cognitive control has been reported by a variety of human electrophysiological techniques, including magneto encephalography.

What this research provides is evidence that long-term meditation was not just a state of mind during practice and that it was more of a learned trait that created long lasting neural-plastic changes in cortical-thalamic circuits, in individuals that were long-term daily meditators. What this indicates is that gamma waves within the brain are linked to volitional attention and the greater the time spent paying attention to one's own psychological and body states the greater the benefits in neural plasticity.

It has been found that the gamma waves originate in the thalamus, they then move over the brain from front to back at a rate of 40 times per second, as they move they collect different neural circuits and bring them into precept, as a result they bring the precept into attention. When damage happens to the thalamus the wave stops, resulting in a complete loss of conscious awareness and the individual goes into a coma. With the above research by Ferrarelli (2013) in mind and the research conducted on the Buddhist monks the evidence points to a conclusion that enhanced gamma wave power and the development of mindfulness leads to a heightened sense of consciousness, bliss, and intellectual acuity subsequent to meditation. Meditation is also linked to a number of heath benefits as well as an overall feeling of wellness, and helps to reduce stress and high emotional states. With all this clear evidence, it is no wonder that coaches and teachers are switching onto the benefits of meditation, let alone those involved in the professional field of dealing with violence and aggression. The Dalai Lama has a daily meditative routine, where he sets aside 4 hours per day every morning, if asked he replies that it is hard work.

From the above research it may at first seem that the benefits are quite new in relative terms, however we know that the practice of meditation has been carried on for hundreds of years especially by Buddhist monks and those interested in the health benefits for long term well being. Is there evidence that the benefits for meditation linked to combat has also been known and practiced? There are many warrior classes that point to evidence that far from being a new paradigm, combat meditation is as old as combat itself, the Samaria and the Native Indian are two such classes, later I will discuss the warrior in more detail.

4 THE BODY SEEKS SYMMETRY

"Learning coordination is a matter of training the nervous system and not a question of training the muscles. The transition from totally uncoordinated muscular effort to skill of the highest perfection is a process of developing the connections in the nervous system" Bruce Lee (1975)

Symmetry is possibly one of the most important of actions that occur within the human body, once we really understand the benefits of this natural heritable process of movement it will enable an individual to move efficiently and effectively.

I intend to explore how symmetry works and where it can be found. There are body responses that do not require symmetry to ensure their speed and effectiveness, what I am referring to here is the inbuilt startle reflexes that the body uses to protect itself against impending danger, pain and other types of stimulus.

An object that is symmetrical has the property of being symmetrical about a vertical plane, however we also have other various types of symmetry. Radial symmetry is symmetrical around a central axis. When an organism is radially symmetrical, you could cut from one side of the organism through the centre horizontally to the other side, this cut would produce two equal halves. Bilateral symmetry occurs along the vertical plane (sagital) and is created by a reflection of images on either side of a centre line.

There are five basic types of symmetry at work in the human body, "symmetry of movement", "symmetry of postures", "symmetry of muscle strength activation", "symmetry of control systems", and "symmetry of features". My intention is to explore movement in more detail and provide some evidence as to why this type of movement within the human body is so important, I will also take a look at postures and muscle strength activation as this also has an effect on the efficiency of movement and the biomechanics that drive the human body.

Human movement originates from a communication system between our neuromuscular proprioception senses and the sensory input and output region within the brain, Neural-Muscular Programming (NMP) along with Fixed Action Patterns and startle reflex, all play their part and will be discussed in detail later.

The region in the brain that is responsible for controlling the movement of humans is the primary motor cortex, located in the posterior portion of the frontal lobe. Wikipedia (2013). The Primary Motor Cortex works in association with other motor areas including pre-motor cortex, the supplementary motor area, posterior parietal cortex, and several sub cortical brain regions, to plan and execute movement. The primary motor cortex sends axons down the spinal cord to synapse onto the interneuron circuitry of the spinal cord and also directly onto the alpha motor neurons in the spinal cord which connect to the muscles.

The primary motor cortex contains a rough map of the body, with different body parts controlled by partially overlapping regions of cortex arranged from the toe (at the top of the cerebral hemisphere) to mouth (at the bottom) along a fold in the cortex called the central sulcus. Each cerebral hemisphere contains a map that controls mainly the opposite side of the body. Later I will discuss early research into the mapping of the motor cortex in primates, which led to evidence that the brain is indeed plastic in every account.

Within the primary motor cortex there is a representation of the various different body parts of humans. The arrangements of these representations are called a motor homunculus, Latin for little person. All the human body is represented on this map, including the extremities, parts of the torso, all areas of the head down to the tips of the fingers, and these, along with fixed action patterns like raising the arm up to grasp an object, all have their place in the homunculus.

The arm and hand motor area is in comparison to the leg, larger in its occupied land-space, and occupies the part of perceptual gyros between the leg and face area. The area that represents the hand and some face parts are larger than any other, with more neurons being assigned to activate and receive incoming stimuli from these areas.

Research has shown that after amputation for example; the area previously assigned to the limb that has been amputated shifts to take up sensory input from another area. The assignment of large areas of the motor cortex to various body actions help us understand why humans have such dexterity in their arms, hands and fingers. Remember the research into taxi drivers in London; their amygdala had grown in size, reassigning more neurons to the activity of remembering the complex road system in London. Using both arms together for a dedicated activity and matching the pattern of movement would over a long period of time produce the same results, larger areas dedicated to such movement, indicating that bilateral symmetry takes up more land space within the primary motor cortex and that the brain is plastic and able to reassign more neurons to a particular activity.

In most cases, symmetry comes naturally without having to consciously think about it. It's when we take it out of our subconscious thought process and apply it to conscious thought that we are able to make extraordinary improvements in the way we move. It has a direct effect on speed, power, alignment principles and many more areas within the combative and martial art arena. An important aspect about adapting new or existing pathways of muscle movement is to know why the human body moves in a particular manner. Understanding this movement is the first step towards more efficient and effective movement, once you have taken new improvements on board and adaptation has occurred, your neuromuscular pathways will start to embed the specific movements, working towards being able to assign them back to the domain from where they came from, your subconscious. Here you will access them without conscious thought and your speed and power will increase substantially. For this process to work at its best we need to be able to really understand what is happening, why we do certain movements and what works and what does not.

Symmetry training of specific movements has a significant effect on the recovery of limb movement after injury Joseph Zeni. Jr (2013) and his associates of the university of Dalaware conducted an analysis on a longitudinal basis and researched the feasibility and effectiveness of an outpatient rehabilitation protocol that included movement symmetry biofeedback on functional and biomechanical outcomes after. Total Knee Arthroplasty (TKA). This involves a surgical procedure in which damaged parts of the knee joint are replaced with artificial parts, muscles and ligaments around the knee are separated to expose the inside of the joint. The ends of the thigh bone (femur) and the shin bone (tibia) are removed as is often the underside of the kneecap (patella). The artificial parts are then cemented into place. The new knee typically has a metal shell on the end of the femur, and the same metal or plastic trough onto the tibia, and sometimes a plastic button in the kneecap. This surgery has resulted in patients experiencing a loss of strength in the recovering knee that has resulted in movement that is abnormal and even after rehabilitation of the operated knee, problems have persisted, resulting in an asymmetrical increase in load onto the knee that has not been operated on.

The method used by Zeni and his colleagues to assess the feasibility of symmetry movement training, was to use biomechanical and functional metrics to assess participants 2 to 3 weeks prior to TKA, then again on being discharged from outpatient physical therapy and finally 6 months after surgery. They assessed 9 men and 2 women all of whom underwent 6 to 8 weeks of outpatient physical therapy that included specialized symmetry training. They compared the 6-month outcomes with a control group that were matched by age, body mass index and sex, 9 men and 2 women, these patients received the normal 6 to 8 weeks of physical therapy but not the specialized symmetry training. Their results were significant, out of the 11 patients that received the specialised symmetry training, 9 demonstrated clinically meaningful improvements that exceeded the minimal detectable change for all performance-based functional tests at the 6-month period after surgery.

These patients had greater knee extension during mid-stance walking; the knee movements were more symmetrical, biphasic and were more representative of a normal knee movement than the patients that did not have the specialised symmetry training. They concluded that the additional symmetry training post-operation was safe and viable to regaining normal symmetrical movement. What this study provides is evidence that the body seeks symmetry in movement and that specialised training can produce clinically better movement after damage. If that is the case, then it stands to reason that when designing movement that is combat based. specialised symmetry training should be an important consideration. It is vitally important that the correct body mechanics are adhered to; teaching movement that is un-natural is one of the key mistakes when efficiency and effectiveness are required.

In a great many martial/combative classes today the words "we will teach you what comes naturally", are all too often heard, the question to your instructor should be; what is natural movement and how do we know it's natural? Observe a newborn child a few days old, when a parent places their finger onto the baby's palm, you see the baby grasp the finger tightly. Be careful though, because the baby cannot control this reflex. If you place a rattle in your baby's hand, for example, they may let go unexpectedly and drop it on their head. A baby's grip is so strong; you may be able to pull them up when they are gripping both your fingers.

This reflex is also present in the feet, causing the toes to curl. It can be tested by lightly touching a baby's feet or toes. This reflex only lasts until a child is about 3 months old at which time it will start to become extinct, the reflex is known as: Darwinian Reflex (after Scientist Charles Darwin), or Tonic Grasp Reflex or Palmar/Plantar Grasp Reflex. As soon as a newborn is able to stand, we see the beginnings of their attempt to walk! None of this is taught, it's simply natural. This theory of development is linked to what is call 'Nativist theory', which considers development as being driven genetically throughout life, creating a process of natural maturation along a predefined path, this is the starting point for what I call Neuromuscular Programming (NMP) which is discussed later.

Finally the one system I have not mentioned is symmetry of control, by this I mean the bilateral control that occurs via both sides of the cerebrum. The hemispheres of the cerebrum are specialized for different tasks. The left hemisphere is regarded as the verbal and logical brain, and the right hemisphere is thought to govern creativity among other things, spatial relations, face recognition and emotions. Also, the right hemisphere controls the left hand and the left hemisphere controls the right hand, bilateral control.

Left vs Right Hand.

Mickevicien, Motiejunaite, Karanauskiene, Skurvydas, Vizbaraite, Krutulyte and Rimdeikiene (2001), conducted research into gender dependent Bimanual task performance, the traditional view is that the right hemisphere is more involved in spatial activities e.g., producing smaller movement errors, whereas the left hemisphere is more involved in temporal activities.

It is understood that the left hand, which is not dominant, is usually employed for movement preparation and the visual-spatial aspects of movement, shorter Reaction Time (RT) to a target has evidenced this. The right hand dominant is usually used in the coordinated timing of pointing actions and more effective use of limb dynamics for the coordination of reaching movement. Movements that require uni-manual aiming are considered to be faster and are more accurately controlled by the left non-dominant hand rather than the right dominant hand. This results in the left hand being more accurate and having a quicker RT. The researchers were more interested and focused on reaction times between genders than a study of symmetry, however the evidence remains that symmetrical movement does occur, even if there are small discrepancies in the RT of hands.

What the above researchers did find, was that in most people a dominant (and faster) right hand implies a dominant left hemisphere. However there was a minority (20%-25%) of right-handed people who actually had a dominant right hemisphere, and reaction time on the right side of the body was slower in these people because commands had to originate in the right hemisphere and then cross over to the left hemisphere, and then get to the right hand. In other words, the side of the body with the longer reaction time (not always the side with the non preferred hand) is the side with the dominant hemisphere.

Miller and Van Nes (2007) cited by Mickevicien, Motiejunaite, Karanauskiene, Skurvydas, Vizbaraite, Krutulyte and Rimdeikiene (2001), found that responses involving both hands were faster when the stimulus was presented to both hemispheres of the brain simultaneously. Because the right (emotional) hemisphere is supplied with input by the left eye, it might be suspected that the left visual field would be the fastest at identifying emotions, evidence that symmetry has a very important part to play across both hemispheres.

Before I look at natural movement I need to provide a clearer understanding of learned maladaptive behaviour ("LMB").

Learned Maladaptive behaviour

As humans, we have the ability to learn and move in a wide variety of ways, some of which are unnatural. For example, break-dancing or even something as simple as walking in a straight line, analyse your own walking gait! How many walk with feet facing outwards? Over time what has happened, is that the major muscles of the leg have become lazy and allowed the feet to fall outwards, this then transfers to the feet and subsequently your own walking gait. This is not so much LMB as it is falling into lazy biomechanics. Remember the learned behaviour of the gun draw? Draw, double tap and return the gun to the holster that was definitely a LMB.

From our early beginnings the one thing man has been doing is fighting, it's part of our make up, and when we first walked the earth, the likelihood is that we never had complicated fighting systems, they came along much later. At a very base level the natural weapons that we have are our teeth, fingers and nails. Over all the years that the Martial arts have been around, different cultures and influences have changed the way we look at forming our natural weapons. The point here is that we developed weapons to fight with, we then termed our own body parts as natural weapons, to make a clear distinction between man-made weapons and natural weapons that could be formed using our body.

From this grew a great many variations of different weapons and the way they can be used. Some of these are simply not natural in any way, the weapon is not natural and the way it's delivered is not either. This does not mean the weapon or part of the body being used as a weapon is not effective, it may well be the case that it is, it's just not natural. Any part of the body could be described as a natural weapon, if it's part of our body and can be used in a manner to cause harm or injury to another human being, then its natural! So to be very specific as to what is and what is not a natural weapon I need to make a clear definition.

Being taught to form a weapon and create a posture with a body part that is not natural creates the difference between natural or un-natural. An example of an un-natural weapon formation within the fighting arts is what is sometimes called a 'Heel palm'. The posture that is un-natural is when the fingers are all touching, curled at the second finger joint, with the thumb tucked in towards the palm of the hand. Although this posture can be learned relatively quickly and used with effectiveness it is a learned posture and not one that we find in our day-to-day lives. Understanding this will help us in identifying what could be termed as natural or un-natural. Postures that are natural are found throughout our day-to-day movement. They are part of our body yes, just not a natural weapon. The method of delivery is also critical in understanding natural movement and weapons. To enable an understanding of natural human movement that follows the principle of bilateral or lateral symmetry it is important that we get to the bottom of this description, it is for this reason that I want to create some further detail here.

Let's look at one very specific weapon and try to discover its effectiveness. Remember nothing is right or wrong, what I am discussing here is natural movement and natural weapons. These weapons and the method of delivery are either, not effective, effective, more effective or most effective Chapel (1991); the traditional Hand Sword is a good place to start. Touching the fingers together and straightening the fingers of the hand is the normal traditional method of forming this weapon, so that it looks like a sword, hence the name 'Hand Sword' right! Choose a target or something that you can hit, and try not to damage your hand. While undertaking a strike remember that it is not what others may feel, it's about what you feel that matters, how effective does the posture feel when impact occurs? Hand Sword the target a few times with the hand formed as described above. Now spread your metacarpals, the ones on the back of your hand, the thumb also needs to be pulled away from the hand, this is a complete reverse of the first posture, try hitting your target with this weapon. Hopefully you will discover that the hand has more structure to it. The posture that you have now created by spreading your metacarpals and pulling the thumb back, is better prepared to take impact when the weapon hits your target, you may well feel that you can hit even harder than before, you could say that this is a more 'NATURAL' posture for the action that you have intended it for.

Next is the method of execution, how do you move your arm and use this particular weapon? There are three basic methods of delivery. 1, use the bicep as the prime muscle behind the execution of the strike. 2 keep the arm at a fixed angle and use rotation from the shoulder as the main method of delivery and 3 use the shoulder rotation as in 2 above and while executing the circular rotation of the shoulder, also engage the bicep to help deliver the strike. The main concept behind this chapter is to discover what constitutes natural movement of the human body, while also understanding that bilateral symmetry is a prime mechanism that allows the body to function efficiently and effectively.

All three methods above involve natural body motion, but only one allows for easy symmetrical movement that does not include a volitional effort of mind to control and this is the important part, as natural efficient movement is always a result of bilateral symmetry. As identified in the research above, specialised symmetry movement training quickly helps to rewire the neural connections within the brain allowing for the body to recover it's normal mobility. Using both sides of the body, rotating both arms from the shoulder at the same time while maintaining a fixed arm position allows for efficient movement that does not have an interrupting thought process of having to bend the arm as well, and creates a faster firing sequence, this in turn allows the body to move faster.

Natural Movement

We all move completely naturally in our everyday lives, it's just what we do, we are very seldom taught how to move or how to breathe, why then is it that when we enter a Martial Arts school or any combat applied system that individuals are taught to move in ways that they have never done before? Movement is so natural that most of us take it for granted, just walking down the street is an example of natural movement, nobody teaches us how to walk! However when we observe people walking, we soon learn that not all of us walk the same.

Take a natural walk for example and compare it with a sliding step often found within a Martial Art. We have never walked this way before and are now being told to slide the foot along the floor to feel where we are stepping! What could be the explanation for such a radical change in how humans move, it's been thousands of years since the species homeo sapiens became bi-pedal and walked on two legs, somewhere in the development of human movement we lost the ability to slide our feet! I have heard explanations like, it teaches you to feel the floor in case something is there; it puts you in touch with the ground! Humans never move within our normal daily lives in this manner but because some high-ranking black belt or instructor tells you to move this way, then it has to be right! No it's not.

Take some time to study the way you walk, do your feet point out to the side, are they straight or do your toes seem to come in towards each other when you walk naturally? Each of the above has varying degrees of angles that the feet create during the activity of walking. Humans are all made to walk in a straight line and usually they place their heel to the floor first then the ball of the foot and transition through the toes as they lift their foot again, this process can be broken down into different stages of a walking cycle; (a), the support or stance phase and (b), the swing phase. Within the chapter on Neuro Muscular Programming I have gone into greater detail regarding this process as the feet are one of the major proprioceptive sense monitors within the human body, however here I am more concerned with the gait and symmetry. Walking naturally requires the foot to maintain an alignment through the ankle, knee and hip joints, when we observe a person walking with their feet pointing out to the side, what we see is a relaxing of the adductor Longus, and the Couturier muscles.

This relaxing of the legs allows the feet to point outwards and with it the knees and the legs in general, effectively walking on the outside of the foot, this is lazy behaviour as it is not a consequence of volitional effort to learn a particular type of behaviour. What has occurred here is that an individual has lost natural movement through LMB, we walk in a manner that is not structurally correct. Over time this will have a detrimental effect on our pelvic structure and the lower joints of the leg. For natural movement to occur we must pay attention to this and make an effort to get our lower platform and legs working with the correct alignment.

All of this will ultimately affect the way we move, making it more efficient; it will also help greatly with a Martial application as well as creating a healthier method of movement. Tyldesley and Grieve (1996) in their book Muscles, Nerves and Movement, Kinesiology in Daily Living, describe an abnormal gait and the possible causes that may affect this, which include mechanical or neurological origins, such as the Total Knee Arthroplasty operation above or a stroke that affects movement of one side of the body, there are many ways in which humans can be affected that cause an abnormal gait. However they also state that walking demands the ability to perform the swing and support movements of the lower limb, at the same time as maintaining the balance of the body.

Co-ordination of the bilateral movements of the lower limbs, torso and arms are essential. The postural reflexes which are largely based on the brain stem, control the balance reactions, such as the torso moving laterally over the supporting leg at each step and the opposite arm counter swinging back as the swinging leg plants. What we find here is that bilateral symmetrical balance is an innate process that does not require cognitive thought processes within the brain, if this were the case our walking would probably be more akin to a stumbling young infant as they began to code in the balance and symmetry required to stand and walk.

Abnormal gaits can be caused by a failing in the skeletal structure, either through bone fractures, or osteoporosis of the bone. Any pain caused by muscle or nerve damage will also affect the normal movement of the limbs. One of the least know effects on gait are the neurological causes according to Tyldesley and Grieve (1996) gait could be affected by

1. Abnormal muscle tone – hypotonia, spasticity or rigidity.
2. Presence of abnormal synergy in the lower limb.
3. Disturbance of postural reflexes.
4. Absence of sensory feedback from the sole of the foot and from proprioceptors in the muscles and joints.

5. Loss of body image when one side is ignored and the affected side is left behind.
6. Perceptual problems leading to difficulty in judging distances and depths and therefore where to put limbs.

These are major problems that can cause our normal bilateral symmetry to go out of phase and ultimately have a devastating affect on mobility. Even small problems with the feet can affect the normal movement of walking, remember when you last stubbed your toe or stepped on something sharp? for a while your normal bilateral symmetrical gait goes out of phase until the pain has subsided. Before moving on, I want to point out how important the mind is in controlling and regulating something as mechanical as a normal walk with an up-right posture. This links itself back to the first chapter where I discussed attention and neural plasticity.

Walking locomotion is by far one of the most important bilateral symmetrical movements that humans do, before leaving this subject let's have a look at some of the abnormal gaits that people have. The following descriptions of gaits and their effect on individuals are from Stanford School of Medicine (2013).

Hemiplegic Gait

The patient stands with unilateral weakness on the affected side, arm flexed, adducted and internally rotated. Leg on same side is in extension with plantar flexion of the foot and toes. When walking, the patient will hold his or her arm to one side and drags his or her affected leg in a semicircle (circumduction) due to weakness of distal muscles (foot drop) and extensor hypertonia in lower limb. This is most commonly seen in stroke. With mild hemi paresis, loss of normal arm swing and slight circumduction may be the only abnormalities.

Diplegic Gait

Patients have involvement on both sides with spasticity in lower extremities worse than upper extremities. The patient walks with an abnormally narrow base, dragging both legs and scraping the toes. This gait is seen in bilateral per ventricular lesions, such as those seen in cerebral palsy. There is also characteristic extreme tightness of hip adductors, which can cause legs to cross the midline referred to as a scissors gait. In countries with adequate medical care, patients with cerebral palsy may have hip adductor release surgery to minimize scissoring.

Neuropathic Gait (Steppage Gait, Equine Gait)

Seen in patients with foot drop (weakness of foot dorsiflexion), the cause of this gait is due to an attempt to lift the leg high enough during walking so that the foot does not drag on the floor. If unilateral, causes include peroneal nerve palsy and L5 radiculopathy. If bilateral, causes include amyotrophic lateral sclerosis, Charcot-Marie-Tooth disease and other peripheral neuropathies including those associated with uncontrolled diabetes.

Myopathic Gait (Waddling Gait)

Hip girdle muscles are responsible for keeping the pelvis level when walking. If you have weakness on one side, this will lead to a drop in the pelvis on the contra lateral side of the pelvis while walking (Trendelenburg sign). With bilateral weakness, you will have dropping of the pelvis on both sides during walking leading to waddling. This gait is seen in-patient with myopathies, such as muscular dystrophy.

Parkinsonian Gait

In this gait, the patient will have rigidity and bad kinesis. He or she will be stooped with the head and neck forward, with flexion at the knees. The whole upper extremity is also in flexion with the fingers usually extended. The patient walks with slow little steps known as "marche a petits pas" (walk of little steps). Patient may also have difficulty initiating steps. The patient may show an involuntary inclination to take accelerating steps, known as festination. This gait is seen in Parkinson's disease or any other condition causing Parkinsonism, such as side effects from drugs.

Choreiform Gait (Hyperkinetic Gait)

This gait is seen with certain basal ganglia disorders including Sydenham's chorea, Huntington's disease and other forms of chorea, athetosis or dystonia. The patient will display irregular, jerky, involuntary movements in all extremities. Walking may accentuate their baseline movement disorder.

Ataxic Gait (Cerebella)

Most commonly seen in cerebella disease, this gait is described as clumsy, staggering movements with a wide-based gait. While standing still, the patient's body may swagger back and forth and from side to side, known as titubation. Patients will not be able to walk from heel to toe or in a straight line. The gait of acute alcohol intoxication will resemble the gait of cerebella disease. Patients with more truncal instability are more likely to have midline cerebella disease at the vermis.

Sensory Gait

As our feet touch the ground, we receive proprioceptive information to tell us their location. The sensory ataxic gait occurs when there is loss of this proprioceptive input. In an effort to know when the feet land and their location, the patient will slam the foot hard onto the ground in order to sense it. A key to this gait involves its exacerbation when patients cannot see their feet (i.e. in the dark). This gait is also sometimes referred to as a stomping gait since patients may lift their legs very high to hit the ground hard. This gait can be seen in disorders of the dorsal columns (B12 deficiency or tabes dorsalis) or in diseases affecting the peripheral nerves (uncontrolled diabetes). In its severe form, this gait can cause an ataxia that resembles the cerebella ataxic gait.

What information is gained from this is that sensory input is critical, once damage occurs, the normally reliable bilateral symmetrical movement goes out of phase, in some cases permanently. However hope is at hand with the more recent understanding of neural plasticity and the effects of specialised symmetry training, improvements in re-wiring sensory information has occurred. To emphasize this I will provide evidence to the major steps forward in Activity Dependent Cortical Reorganisation.

Activity Dependent Cortical Reorganization

The road to Activity Dependent Cortical Reorganization (ADCR) has certainly been a long one and with it the understanding of bilateral symmetry, its effects on remapping the brain, how our habitual habits affect movement and the efficiency of such action, causing the reorganisation of neural pathways.

'Neural plasticity' started way back in the early 1900 with William James (1890) he writes " if habits are due to the plasticity of materials to outward agents, we can immediately see to what outward influences, if to any, the brain-matter is plastic. Not to mechanical pressures, not to thermal changes, not to any of the forces to which all the other organs of our body are exposed; for nature has carefully shut up our brain and spinal cord in bony boxes, where no influences of this sort can get to them. She has floated them in fluid so that only the severest shocks can give them a concussion and blanketed and wrapped them about in an altogether exceptional way.

The only impressions that can be made upon them are through the blood, on the one hand, and through the sensory nerve roots, on the other; and it is to the infinitely attenuated currents that pour in through these latter channels that the hemispherical cortex shows itself to be so peculiarly susceptible". He touches on the fact that actions, once created and a neural path activated, with a subsequent action along the same path by a nerve current, would further make the pathway more permanent than before.

I like the analogy of a freshly grown field of grass blowing freely in the wind, one day you decide to walk the same field every day to enjoy the views and the peaceful times it gives you, you make a conscious effort to everyday walk a different path. After a while you look back on the path and see no trace that you have ever been there. Then one day you decide to make a volitional effort to walk the same path every day, after a short time you notice a path beginning to show, you find it easier to follow the path now. After a little more time a track has been forged now and you do not have to even think about the mechanics of walking the same path every day.

These tracks within the brain are cortical representations, they are not inherited pathways already sown into the blueprint of our DNA, and they are never the same, much like each individual's fingerprints. Research conducted on monkeys in 1912 by Charles Sherrington and T J Brown identified that each monkey's brain, when stimulated by an electrical impulse, resulted in a different map, an impulse in one monkey stimulated a sensation on the palm of the hand, the exact same spot on another, the left cheek on another, a different muscle in a different spot and so on. This confirmed at the time that cortical organization was dependent upon the learning experiences and the historical use of the motor system of individual monkeys, these were the tracks across the freshly sown field of grass and no map was the same.

After the research of Sherrington and Brown came Karl Lashley (1890–1958), he was a distinguished psychologist of the 20th century, known for work in physiological and comparative psychology. Lashley was a materialist he believed that the origins of behaviour could be studied within the brain. There was a possibility that the research carried out on mapping of monkey brains may have been flawed, as the individual differences may well have been result-inherited experiences. Lashley's research involved mapping a monkey's brain over an extended period of time to discover if cortical reorganisation occurred within a short time scale, would the brain shift neural activity dependent upon usage? Over the period of a month he mapped the brain of a single monkey and compared the maps to one another. If it were indeed the case that maps were inherited then there would be no difference in the map over the course of a month, let alone a year. He discovered each map that was taken differed in detail and when compared over longer times the change was even more dramatic.

This was evidence that the experiences of each monkey re-wrote the cortical regions of the brain, reflecting the experiences of the monkey; activity dependent cortical reorganization was a real process that indicated that the brain was plastic. Lashley also concluded that a larger area of cortical representation 'land space' was used when muscle activity in a particular area was repetitively used. This is a very significant statement when it comes to bilateral symmetry being natural movement, what this conclusion is saying is 'that the more a primate uses a muscle or set of muscles the more land space is attributed to that movement' this is possibly the first explanation to muscle memory!

Every time that you walk across that field of grass the neural activity that controls your walk grows, becomes ingrained into the deep pastures of your brain and when that day happens when you start to walk another path or stop altogether, then the grass grows over that once well trodden path and so it is with human movement. If behaviour changes, the motor cortex reorganises the neural activity and this is one compelling reason why humans have that all too great an ability to become lazy, to learn maladaptive behaviours and once ingrained within our motor cortex, believe that the behaviour that we now do within our given discipline is the most natural, efficient and effective means of achieving our physical goals.

So what do I mean when I use the term "the body seeks symmetry"? Clearly what fires together wires together, we therefore must not forget that we have two sides of our body that are controlled by different regions within our brains, therefore we should be mindful of creating an imbalance in cortical representations.

Within the bodybuilding world there is a term 'Unilateral movement', this is used in conjunction with bilateral movement, there is no conflict, just a different understanding for the reasons behind unilateral movement. The train of thought regarding strength training and these two different methods is that bilateral symmetry is good for maximising the strength building of muscles, however the dominant limb has the capability to take more of the load, creating an imbalance in the tone and strength of the muscle.

It is for this reason that unilateral training should also be used in tandem with bilateral movement. The unilateral training will have the effect of correcting muscle imbalances, while at the same time improving bilateral movement. In terms of ADCR this will create an imbalance in the land occupied by the different actions upon the cortex of the brain. To be efficient, mirroring needs to occur and this can only be effectively achieved when bilateral symmetry is employed.

Take an Olympic sprinter as an example and ignoring the mechanics that are needed to launch the body into the sprint from a standing start, let's look at the action of the sprint itself. Mentally picture the last time that you observed an athlete in full flight, what does their body mechanics teach us? Well the first thing that should strike you is how their upper body limbs are helping to drive the legs. How, when the right leg is powering forward, the left arm is doing the same movement, it's a reverse of what's happening on the ground, when the left leg is finishing its stride behind them, the right arm is doing the same.

Everything is working symmetrically and in harmony to power the body forward. All these mechanics are coming together sub consciously, if they were to have to think about each action they would soon lose their co-ordination, speed and power. Imagine if you were to tie the right arm of a race winner to their side and then repeat the race, or even use unilateral training to practice each side independently of each other, the sprinter would not be able to work at their optimum efficiency, the body would be out of sync, it would not be working symmetrically and it would soon become apparent that they were not capable of obtaining the same high level of performance, their movement would be un-natural.

How does this all translate to movement within the combative martial arts arena? Let's set up an experiment, remember that the process needs to be repeated accurately each time to get the full understanding and the feeling of what we are trying to achieve. Raise your right arm up to a horizontal plain, no higher than your shoulder, imagine a symmetrical box, which is formed from left shoulder to your right shoulder, out in front of your right shoulder across in front of you then back to your left shoulder, with your arm raised ensuring that your fist does not change its position from its natural alignment. Have the fist slightly out in front of the box, so that your arm has formed an obtuse angle. Have your left hand open and positioned facing forward at the same height as your left shoulder.

Now with your right hand execute a back fist strike, as fast as you can striking a target that is in line with your right shoulder at the end of your reach with your arm, do not over extend or lean. Execute this with your left hand still, now once this has been done, attempt to execute the same strike, but this time move your left hand to your right shoulder slowly, when the left hand touches the shoulder; ensure that the hand is high enough so that the fingers slightly overlap the shoulder. What you should find here is that it is almost impossible for you to carry out the prescribed movement in the way described, you simply cannot move one fast and the other slow! At least not without a great deal of cortical re-organisation and practice. Of course there is always the possibility that you could learn this type of behaviour over time, it's not natural though. Now the last part of this experiment, execute both movements as fast as you can, in fact what can help here is not to think about the right hand, think about the left moving as fast as it can from its starting position to your left shoulder and back to its original position, do the same thing with your right hand, moving it from its point of origin out to its target and back again to the original starting position.

Now you should have discovered that this type of Bilateral Symmetry has increased the speed of your movement significantly. If your intention is to execute a strike at speed with efficient body mechanics, it should be just that "Fast" it does not have to have any power principles to back it up, its effectiveness relies on speed, the quicker it moves, the more velocity is generated and therefore more damage will be caused. The greater the velocity the greater the damage!

The lesson here is that without the body moving symmetrically you will not be able to move at the highest velocity you are capable of and in your most effective manner, especially when speed is your desired outcome, it will be like tying the sprinter's arm to his side creating an imbalance in his natural body mechanics.

When you are launching a weapon away from its platform at high-speed, the action that is required to stop your fist mentally is hard, damage can occur at this point by applying the brakes to the move prematurely, that's why it needs a target to hit and dissipate its energy into. The hit is the trigger telling the brain to stop the strike and begin its returning motion. When you move both hands at the same speed the hand hitting the shoulder is the corresponding trigger to stop the left hand, everything works symmetrically. One of the easiest movements to see symmetry in action, is the traditional reverse punch, as one hand is hitting the other is returning. If power is required, the returning arm has to return with power that matches the power of the punching hand and the more practice, the greater the ADCR is.

The key to understanding bilateral symmetry is to know how and what you are trying to achieve. It has to be practiced by accurate repetition, so that eventually you can put it back into your subconscious, when this happens it will become even faster, as you no longer have to rely on conscious thought, which in itself is slower than your sub conscious thought process, you will have created a field of firing neurons that have embedded the action into your brain, you will have muscle memory and the 'land space' within your somatosensory cortex will be large and attended to. The only way that we have a chance of moving at a high-speed is to ensure that we are moving as efficiently and naturally as possible, if our movement is anything else, then we run the risk of being too slow. Speed is not always the desired outcome, there may well be times when speed is redundant to power, and this will ultimately depend upon the circumstances.

Symmetry of Posture

There are two basic differences between postural symmetry, one is concerned with the posture of the body as it relates to our normal everyday activities, like standing or sitting. It's how we hold our posture and can have either a positive or negative effect on our health and mobility. The other is the posture that we create either volitionally or spontaneously and it is this second area that relates to the subject of this chapter and has an impact on the effectiveness of performance and symmetry. But first let's look at the health implications of good or bad symmetrical control of the human body.

From the moment humans are born, they begin a long road to mastering the symmetry of posture and then, for reasons that will become apparent, go into a sharp decline in symmetrical posture. Newborn infants start to develop their posture control very early and progress through different stages, each stage acting as a springboard for the next. For an infant, development of motor autonomy is a prime genetic blueprint driving them on towards complete independence. Although sometimes an infant can skip a stage, in general they follow a defined developmental process, starting from lying to rolling, pushing up, sitting, crawling, standing and then walking. For the infant, each stage brings with it more advanced control of balance, symmetry and posture control. During these stages, reflexive behaviour can also be seen to develop. Once the infant has learnt to stand and walk his posture can be observed in more detail, one very critical element is the erectness of the posture, which continues to maintain itself as the infant develops into a child.

At the beginning of our development the body assumes its optimal posture for standing and locomotion, at this time unless there are genetic problems that cause the body to develop without a symmetrical blueprint, then movement and control will be effective and efficient. There is however one great impediment to maintaining this optimum posture, one in which we all have to suffer and that's the force of gravity. Humans today have a radically different life than they did 100 or 500 years ago, we now, in the majority of the developed World, live a sedentary lifestyle, we drive to and from places, we have developed technology to assist us in almost every way, from escalators that make stairs extinct, to shopping trolleys and mechanical diggers, they all play their part in helping humans move and locomote without effort. This in turn exposes us to an increased chance of injury due to a lack of exercise, muscle strength and practiced efficient movement.

We learn maladaptive behaviours through laziness or listening to an authoritive figure instructing a movement pattern that will lead to bad posture and movement control, compromising our once erect and symmetrical posture, with gravity always there willing to lend a hand to increase the effect upon us. Eventually the body, its posture and symmetrical makeup breaks down leading to ill health and long-term chronic pain. Maintaining good postural symmetry through these testing times is by no means an easy one, those that manage to do it usually do so by luck rather than judgment, have a healthy circulation, are more energetic, their coordination and neuromuscular responses are quicker and they usually have more natural biomechanics, they move easily without pain.

One result of gravity on the human frame over time is a leaning effect, which is usually apparent from the pelvic girdle up through the torso, with the head well out of alignment and tilted forward. Over time, what has occurred, is that the muscles designed to maintain an erect posture have began to give way to the force of gravity affecting the erector spine muscles which hold the body erect. The muscles on the front of the body shorten, resulting in the muscles on the back of the body lengthening, once this occurs the well known stoop becomes visible, creating an imbalance in muscle strength leading to symmetrical failure. It's not only the erect posture that suffers, movement symmetry can be significantly affected. Losing symmetrical control of one region of the body will lead to damage and loss of symmetry in other parts.

When humans move, a dynamic series of alignments occur, all of these have to be in sync and coordinated to allow the body to move efficiently and effectively, if one of the links breaks down then a loss of performance, injury or damage may occur. If it is persistent, long-term chronic pain and ill health may result.

There are individuals and groups that subscribe to specialised symmetry training, the research conducted by Zeni and his colleagues on patients recovering from TKA earlier provided evidence that outpatient rehabilitation protocol that included movement symmetry biofeedback on functional and biomechanical outcomes after the knee operation, significantly aided the individual's recovery back to normal symmetrical body movement. Another group that specialise in this area are Symmetry for Health.

Symmetry for Health use specialised techniques to help them identify any cause of postural weakness or dysfunction that may be responsible for creating some types of mechanical fault or ill health. These techniques include the use of a Palpation Meter, used to measure an individual's specific misalignment that may have become almost undetectable. They also use Symmetry patented software, which uses a three-dimensional view to calculate where the weaknesses are and determine specialised and specific exercises to correct the misalignments. The exercises that are then required to redress the misalignments and imbalances within the body are made specific to that individual, the aim is to re-train the whole body, bringing it back to an aligned symmetrical equilibrium and to further provide life long attention to natural symmetrical movement. In their words "The specific exercises we select (from a database of over 270!) mimics the stimulus we no longer receive through our daily activities. This serves to "re-create" the neural pathways that our ancestors relied on to keep their brain sending correct signals to their muscles to tell their muscles how to correctly hold and move bone" Symmetry for Health (2013). In addition, Symmetry training restores strength and balance to the particular muscle groups that are in need of equalising to ensure they have the physical ability to perform optimally.

A classic example of asymmetrical imbalance in muscle tone and strength, is the infamous "back pain" that in the majority of situations starts as a muscle spasm usually due to our increasing sedentary lifestyle, too much sitting or vegging out on the sofa watching our favorite TV show, which then leads to a muscle imbalance. Our core muscle structure that is designed to hold the body erect and straight becomes a little more inactive every day eventually resulting in over correction of strength in some part of the structure that leads to weaknesses in another. These postural weaknesses are compounded by the relentless efforts of gravity in pulling the structural frame out of symmetrical alignment. Imagine a wheel on a car rotating around an axis on a bearing which when new, is lubricated to ensure ease of use and long lasting wear and tear.

Over time and with a lack of maintenance the bearing begins to dry, then a slight miss-alignment occurs, no correcting repairs are carried out and the problem continually gets worse, eventually resulting in a catastrophic failure of the mechanism, well that's exactly what happens when constant misalignment is allowed to continue in the human mechanical frame. Here I sit typing at my computer becoming aware of my own bad posture that I have, that subtle little ache that has started in my left side!

All of these small asymmetrical imbalances in time starts to affect the overall equilibrium of the body symmetry and leads to postural changes that over time reveal themselves as chronic pain due to damage.

We live in a sedentary society, one that encourages a reliance on technology, automation and mechanisation, one that is now in most parts of the developed world a polar difference from where the human race was a thousand years ago. Pre historic man was designed by the manipulation of evolution over hundreds and hundreds of generations and now today we find ourselves still closely aligned to that ancient blueprint, efficient and effective movement was chosen over and above maladaptive movement, for those genes that did not produce human movement that followed this rule would have been quickly identified as a high cost to the society at the time resulting in early death by a predator or the tribe itself.

We still see this today in our primate relations, when members of the family that cannot pull their weight due to a physical impairment are excluded from the group. Does primate social and natural body movement follow a genetic blueprint? According to Plomin, R. (2001) cited by Chang (2013) genes have a direct correlation on behaviour "The influence of genes on social behaviour is undeniable because genes shape the neural circuits that produce behaviour"

Genes are undoubtedly the universal building blocks for the human and primate species along with every other sentient organism on the planet, this is further supported by Chang (2013) "The adoption of pre-existing biological mechanisms for social purposes, and indeed the evolution of social behaviour in general, must, therefore, have roots in genetic change, or, in more Darwinian terms, must be based on modification through descent of inherited material. One hint that social behaviour influences change in gene pools over time is a handful of studies linking sociality with fitness" This being the case how do the majority of animal species locomote themselves within their environment and is there a genetic blueprint to bilateral symmetry?

The human body has evolved in such a manner that efficient and effective movement has been selected, over and above ineffective or inefficient movement. One of the primary mechanisms for movement is bilateral symmetry. Before I focus on the species homeo sapiens, let's look at research conducted on maneuverability to explain the maintenance of bilateral symmetry in animal evolution.

Holló and Novák (2012) researched locomotion in three-dimensional macro-world space, this in itself they state, is sufficient to explain the maintenance of bilateral symmetry in animal evolution. Their hypothesis is built on an animal's ability to change direction, which is a key element of locomotion that requires the generation of instantaneous " pushing" surfaces, from which the animal can obtain the necessary force to depart in the new direction. Their research shows that "bilateral is the only type of symmetry that can maximize this force; thus, an actively locomoting bilateral body can have the maximal maneuverability as compared to other symmetry types. This confers an obvious selective advantage on the bilateral animal. Implications of the hypothesis: "these considerations imply the view that animal evolution is a highly channeled process, in which bilateral and radial body symmetries seem to be inevitable". They go on to say that 99% of all animals are bilateral, it's no wonder that symmetry is such an important mechanism in human movement. They focus on the aquatic environment because bilateral symmetry (and animal life itself) formed there, and had to be maintained there for millions of years, before bilaterians conquered the land.

In order for us to get a grips with the relationship between this research and the subject of this chapter let's take an in-depth look at their research, they start with the elementary physical fact that to locomote in a fluid takes effort and symmetry. What follows is an extract from their research; a body has to overcome drag (the resistance of the medium in which the body moves, acting in opposition to the direction of locomotion).

The magnitude of the drag force is: $F = -\acute{o} \rho c A v2$ where F is the drag force, ρ is the density of the medium, c is the dimensionless drag coefficient dependent on the body shape, A is the area of the maximal section of the body in the direction of motion, and v is the body' s velocity. The negative sign on the right side indicates that drag is opposite to the direction of motion. It is important to note that this equation is valid for situations where the viscous forces are negligible compared to inertial forces, in what is loosely described as the macroscopic world (i.e. at high Reynolds numbers).

In the microscopic world, the forces are dominated by the viscosity of the fluid rather than by the inertia (i.e. at low Reynolds numbers), however a discussion of the locomotion in the micro scale world is not the concern of this paper. Given the fact that the medium imposes resistance on the body, if resistance forces are unequally distributed around the body, their resultant force will not be zero compared to the rectilinear direction (i.e. movement straight ahead), so the body will not move on a linear path. This is the case when a moving body is asymmetric.

Thus, it follows that a directionally locomoting animal has to be symmetric in order to avoid this effect. To be able to move forward, the animal can have any type of symmetry, so the approach outlined here is not sufficient to explain the success of bilateral symmetry. Rectilinear motion is, however, not the only element of locomotion. One other important element is changing direction, the importance of which, in this regard, has been mostly ignored in the literature so far. A slight deviation from the straight trajectory can easily be obtained by flawing one element of symmetry, thus generating asymmetry in the original direction of motion.

Any symmetrical body can achieve this. However, when a quick changeover is required, the situation becomes very different.
In quick changes in direction, the body has to exercise a force in the opposite direction to the desired new orientation. This means that it has to have a " pushing" surface in water from which to depart in the new direction. This surface is formed by the water layer against which the body is standing, this provides a brace, which in turn allows the body to push itself away, the resulting effect is a great instantaneous drag force. Since ρ in the equation is unchanged, and v is diminishing or constant, the animal has to increase the maximal surface a and/or the drag coefficient c.

We will now overview the main symmetry types in terms of their capacity to create a pushing-off drag force. Given that a swimming body has to minimize the overall drag, its skin friction, and thus its wetted area, has to be adequately reduced. Thus, only three main body forms can be considered: spherical (with endless symmetry planes and symmetry axes), cylindrical (with endless symmetry planes and one symmetry axis) and bilateral (with one plane of symmetry). An elongated radial body that shows a star-like section is suboptimal since it has a very large surface that is far from ideal for swimming forwards.

A spherically symmetrical body cannot generate the pushing surface, being of equal shape and drag in every direction. Since the forces required to change direction are different from the force operating in the direction of its motion – the forces acting on this body are all equalized, it will not be able to depart in a new direction. It can only rotate around itself to deviate to a small extent (as soccer players bend the ball), but this is hardly an effective changeover and obviously cannot guarantee accurate maneuverability (understood simply as the capacity to perform quick and accurate changeovers). In this context, we can disregard how it was able to move directly in the first place.

A cylindrical or approximately cylindrical (or radial) body locomoting with lateral or vertical undulation is able to increase A, which will be generated by a section of its body opposed to the direction in which it wants to move. The area of this surface is given approximately by the product of the diameter and the length of the body portion in question (and of course by its angular orientation to the axis of translation). If its lateral drag coefficient (c) is greater than the frontal, then when the animal turns its body it can also increase c in the equation. However, regardless of the relationship between the anterior and lateral c, if the product (c A) from the lateral view is greater than that from the frontal one, this body will be able to move forward as well as to change direction.

Bilateral symmetry has also proved to be successful both on land and in the air. On land, the force-generating role of the drag in water is replaced by gravitation and so by the necessity of leaning on the land. In this regard, locomotion on land is analogous to that on the fluid solid interface. This locomotion essentially occurs in two dimensions thus, direction shift on land requires the body to be capable of turning left or right, and of being supported from the right as well as from the left.

The effectiveness of creeping locomotion has been improved by the evolution of limbs, which placed on the two sides of the bilateral body, satisfy the above-mentioned condition. (For the sake of simplicity, we will not deal with the limbless evolution of snakes and limbless lizards here.) Flying, similarly to swimming, requires the animal to create pushing surfaces in the air. The evolution of large-surface wings allowed the animals to locomote in a medium, which, compared to water, has a lower density and as a consequence, is almost completely lacking in the hydrostatic pressure that to a certain extent counterbalances the force of gravity in water. The combination of bilateralism with the centralization of the nervous system and cephalisation allowed the evolution of really successful body plans ensuring precise locomotion and rapid information processing.

Holló and Novák (2012) conclude that, from the moment bilateral symmetry arose in macro-animal evolution it represented a potentially enormous selective advantage over other body plans assuring faster change overs and a more precisely directed locomotion. This is a key to survival both for prey and for predators.

This evidence points to a very clear conclusion that human movement is driven by the process of bilateral symmetry and that, as a result of imbalances in this system, less efficient movement will result.

The human body has been designed and selected according to an evolutionary blueprint, one that selects the most effective movement that allows for survival of the fittest. One of the basic requirements for continued efficient movement that the body requires, is constant stimulus, it has to be exercising and moving in its most natural form, this will enable symmetrical muscle tone to be maintained. Pre historic man lived in an environment that was drastically different from today's cities and rural expanses.

To negotiate life back then, humans had to be able to move efficiently, predation was one of the primary motivators for survival and to do this activity required a more active lifestyle than we have today. We had to move in order to hunt, run away from prey and make no mistake, our past ancestor's movement had to be efficient; in other words, optimal physical function determined our survival. Today's descendants are still wired to this genetic blueprint and as such humans still have, not only the movement designs, but the psychology that enabled us to survive as well. We know that in-activity for any extended period of time due to injury or illness will soon start to affect muscle tone, muscle groups begin to weaken and atrophy sets in, eventually resulting in an inability to move in any natural way.

The body has two muscular systems, intrinsic, designed to hold the body upright and resist the constant force of gravity, this system is the primary group of muscles assigned to holding symmetrical body posture in equilibrium. Essentially, this system holds the body at right angles via the ankle, knee, hip, shoulder and head, these are all aligned with gravity. Dynamic muscles, are the larger muscle groups, designed to move the body through large contractions via muscle groups such as quads, hamstrings and abs. The actual biomechanical movement of the human frame is centered on the body being held at right angles, this then facilitates efficient and effective movement in a symmetrical manner. It's the imbalance created by a failure in certain muscle groups that slowly leads to weakening of large muscle groups. When this happens the body has to compensate for this loss of strength by recruiting muscle groups that are not designed to support the new role, this is the point where asymmetrical posture starts to occur, body posture changes and pain sets in.

Given the above information on how and why the body needs to maintain optimum working efficiency, it's easy to accept that bilateral symmetry is genetically inherited and has a direct effect on the performance of the human body. Any coach or instructor that requires the most natural of movement should understand this important area of biomechanical movement, instructors need to pay attention to this and not continue to teach movement that has no symmetrical basis.

Symmetrical Mirror Postures

So far, it's been a road of discovery as far as bilateral symmetry is concerned, however before I leave this subject let's take a look at mirror postures and no I do not mean what you see reflected back at you when you first get up in the morning!

Mirroring occurs bilaterally through the postures that are created and non-cognitive reactions that mirror themselves on each side of the body. The first place to start is mirroring of postures. Reflexive actions create instantaneous mirror postures; an example would be your hands being drawn to protect the face. Imagine that you are standing behind a window that gets hit by a stone and shatters, the inbuilt startle reflex will close both eyes and begin to move the head away from the breaking glass. At the same time your hands will come up to your face, palms facing outward, all of these actions happen in a microsecond. You have no time to create different postures with the hands, they will both mirror themselves, and you will not form one hand as a fist and leave the other open!

In order to understand this a little more, we need a little more detail. Within the brain, co-activation of both hemispheres plays a critical role in generating mirror movements; these hand movements are involuntary muscle contractions as opposed to a volitional effort to form a specific posture with both hands. Mirror movement refers to simultaneous contra-lateral, involuntary, identical movements that accompany voluntary movements. Erlenmeyer first used this term in 1879. The definition of mirror movement as involuntary, synkinetic mirror reversals of an intended movement of opposite side was coined by (Cohen et al. in 1991) cited by Nadkarni (2012) This type of movement is defined as neurologic signs, seen uncommonly in clinical practice.

Research into the relationship between dominant left or right hand use and mirror movements is unclear, a study of hand grip forces was carried out by Uttner (2007) and colleagues, they measured repetitive grip force changes performed at slow and fast frequencies with the active hand and recorded mirror activity in the opposite inactive hand in 17 healthy left-handed (LH) and 17 right-handed (RH) participants. Mirror movements

(MM) are a phenomenon referred to as unintended movements corresponding to homologous muscles in the limb opposite to the one performing voluntary movements (Connolly & Stratton, 1968; Zulch & Muller, 1969) cited by Uttner (2007). A study was undertaken that had participants grasp two identical wireless force transducers between the tips of the thumb and the other fingers in both hands, they were then tasked with increasing the grip repetitively on one hand, while holding the other sensor with the other and not increasing grip strength.

What the data indicated was that mirror activity correlation was more pronounced when the non-dominant hand was active in RH, while LH did not show such asymmetry. It also suggested that the dependence of mirror activity on the active hand in right-handers but not in left-handers reflects the more general finding of stronger manual asymmetry in brain activation as well as in functional performance in right-handers. Involuntary co-activation of the opposite limb during voluntary arm movements is a well-documented phenomenon in healthy adults (Cernacek, 1961; Durwen & Herzog, 1989; Herzog & Durwen, 1992; Linke et al., 1992; Armatas et al., 1994) cited by Uttner (2007).

This supports the fact that sympathetic contractions occur in the muscles of one limb when the other is being stimulated and in doing so, also supports symmetrical hand postures when faced with high stress situations, rather than trying to train one hand to be open and another closed, both should mirror each other if optimum efficiency is your goal.

Research conducted by Heim (2007) with the purpose of identifying environmental and physiological factors that may interact to bring about accidental discharges of firearms; and to make suggestions regarding the training of police officers with the aim of reducing such incidents. They fitted a pistol with a pressure sensor on the hand stock and the trigger and then had the participants undertake 13 different tasks, each of which required the use of the different limbs while still holding the weapon in the opposite hand.

What they found, was that motor activity in different limbs can lead to a significant increase in grip force exerted on a firearm and that the amount of force exerted on the weapon is dependent on the intensity and type of movement and the limb involved in the movement. For movements of the contra lateral arm, a tendency for higher forces to be exerted on a weapon during pulling than during pushing movements was found, whereas the force with which the movement was performed did not seem to have any influence. In contrast, for movements involving the legs, findings indicate that increasingly more forceful leg movements led to a progressively higher risk of unintentionally discharging a firearm due to unintended muscle activity, whereas the type of movement does not seem to influence the amount of force produced in the hand carrying the weapon.

Generally, the use of the lower limbs appears to offer a greater danger for involuntary discharges resulting from unintended muscle activity than movements involving the contra lateral arm. This research provides one of the first bodies of evidence that support bilateral symmetrical movement as being hard wired into the blueprint of humans, which has the effect of creating unintended muscle activity in the hand carrying a firearm, which may evoke involuntary discharges.

What this chapter has provided is evidence of the importance of bilateral symmetry when it comes to training a person to move in their most effective and efficient manner. All too often what will happen in high stress situations is that the individual will return to their dominant movement above all other that may have been lightly trained in. To ensure that any training has been thoroughly coded into the brain will take a very large amount of repetitive actions, performed under the same conditions, or as close as possible to a real situation. Activity dependent cortical reorganisation will happen and the best way to help the encoding process is to pay attention to every detail, ensuring that maladaptive behaviour is avoided. Symmetry plays a significant part in this process and humans are hard wired with the capacity to undertake such actions.

5 WARRIOR MINDSET

"What we do in life echoes in eternity" From Ridley Scott classic film, The Gladiator (2002). Spoken by the film's hero, Maximus who plays the Gladiator.

For thousands of years mankind has depicted a warrior as a solider, part of an army ready to kill an enemy. In today's society the warrior has taken on a different persona, this chapter takes a look back at a couple of warrior societies of our past history, delving into the traditions and mindset of those individuals. I will also discuss the evolutionary process involved in why we have developed such violent individuals. In the past a description of a "Warrior" would be individuals whose time was spent in the art of warfare, they engaged in the actual physical activity of fighting, were the blunt end of the stick or the sharp end depending upon the context. A warrior can be a soldier or an individual person who fights for their brothers and sisters in arms, their family or country, they fight for glory, to a warrior, the journey throughout life becomes a way of life at the exclusion of most other distractions.

In the past history of mankind, such classes of warrior would have been, Japanese Samurai, American Indians, Roman Gladiators, Medieval Knights or African Zulu. In today's society, it's an army solider or Special Forces operative, this then pervades down to individuals that go into harms way to support the law of the land or fighters that put themselves into dangerous competitions between other fighters, MMA or ring fighters for example.

What sets a warrior apart from a combatant that enters the arena for selfish reasons is the way that they live by a certain moral code, they fight in a more disciplined way, their lives are dominated by personal values of honour and a certain morality is applied to the whole mindset of war, an example of this would be in the film "300" when King Leonidas commands his men "prepare for Glory" this depicts where a warrior has to be when engaging in deadly combat.

To demonstrate the difference between an individual who enters the arena for selfish personal goals as opposed to an individual's self-awareness of morality and honour would take more time than this chapter will allow, however you will gain a sense of what that difference is.

In the majority of human cultures the mantle and responsibility to fight falls on the male of the species, quite often the whole of a particular society are involved with the transition from boy to man and most males are capable of becoming potential fighters although not all do. That is not to say that females are not, there are some notable exceptions to the norm.

One such exception to this rule is the women warriors of Dahomey, an eighteenth and nineteenth century Western African kingdom. They were called 'small black Sparta' as they were known to share the same values and military prowess as the Spartans. The women of this warrior culture were raised from birth on a ruthless discipline of physical exercise and combat tactics, they were taught to hunt and became skilled in weapon use. They endured hardship and were also known for their ability to play music and dance. These were women soldiers in the service of the West African kings of Dahomey.

In 1845 Dahomey had a standing army of 12000 soldiers, 5000 of whom were women. Consider this example of the Amazon warrior culture as described by Alpern (1998) in his book The Women Warriors of Dahomey "The amazons renounced marriage as servitude. When they were not fighting, they farmed the land and tended to livestock. For protection they took two months off every year to couple with men from a neighbouring tribe, at random and in the dark. Only women that had killed a man in battle could give up their virginity. Girls born from such unions were reared as amazons; whereas boys had various fates – some were returned to their fathers community; some were deliberately crippled and kept as slaves to do such menial jobs such as spinning wool; some were sacrificed, presumably to the gods of war" For these warrior women, their purpose in life was to make war, they lusted for battle and entered it with cold blood determination, they fought with valor and from the outside were without fear. According to Alpern, in victory they were pitiless, terrifying their neighbours, men regarded them as worthy adversaries The warring fighters were, not all women, however they were commanded by the females, the Amazon male fighters served under the females, who could outshoot and outfight the males. Whatever our view on the role that women play in today's society, clearly there have been occasions in history where men were not always the dominant exponents of war.

There have also been individuals that have become known as warrior women, who have what it takes to take on man at his own game. These are sprinkled throughout history in myth and legend with some historical evidence to support the stories; let's take a quick look at two such women.

Queen Boudicca

Boudicca was the wife of Prasutagus; the ruler of the Iceni tribe that lived in a region of England now called East Anglia. After the Roman invasion of England in 43AD, they made pacts with local tribe leaders to maintain control over their lands. They were allowed to continue ruling their people in exchange for peace and stability. It was Roman law to allow allied kingdoms their independence only for the lifetime of the king who ruled prior to invasion. He would then agree to bequeath his kingdom to Rome in his will. Roman law also allowed inheritance only through the male line, so when Prasutagus died, his attempts to preserve his line were ignored and his kingdom was annexed as if it had been conquered, at which time the Romans took possession of his lands and people, they are credited with having beaten Boudicca and raping their two daughters. While the Roman governor Paulinus was in North Wales engaged in suppressing the Welsh, Boudicca raised an army that consisted of several local tribes numbering around 120,000 men and rebelled against their Roman rulers. As a woman, Boudicca was resolved to win or die; if the men wanted to live in slavery, that was their choice, she had decided to fight at all cost, testament to the will of a warrior. They sacked Camulodunum now Colchester and defeated the Roman IX Legion. They then continued on to destroy both Continuum (London) and Verulamium (St Albans). By the time Paulinus heard of the rout and return from Wales, Boudicca had destroyed much of the ruling Roman positions in the South and taken 70,000 Roman lives in the process. The Roman army in Britain regrouped in the Midlands and finally defeated

Boudicca and her army in the Battle of Witling Street. It is said that during the final battle between the two armies, Boudicca was seen wielding a spear and riding on a chariot into the Roman army, this is depicted in the statue of Boudicca by Thomas Thorneycroft, standing near Westminster Pier, London. How Boudicca died is not known, it has been written that rather than be taken prisoner she and her two daughters drank poison.

"Mad" Ann Bailey

Ann Hennis Trotter Bailey was born in Liverpool England in 1742. Both of her parents died when she was 18 and by 19 she was on a boat sailing for the new lands in America, and later became known as "Mad Ann" for her acts of bravery and heroism that given, the time in history, were considered to be somewhat eccentric for a woman. In 1765 she married Richard Trotter and from this union produced a son who she later gave up at the age of 7 after her husband died, it was around this time that she earned the nickname "Mad Ann", she swore to avenge her husband's death and began wearing men's clothing and taught herself how to shoot a gun. She volunteered her services as a scout and messenger during the Revolutionary War. Bailey's reputation was partly made when the Fort Colonel learned of an impending attack on the Fort and gun powder supplies were low, he asked for a volunteer to ride out for resupplies, when no male stood up she volunteered to ride out into Indian occupied territory, alone and having to cover 100 miles from Fort Clendenin to Fort Savannah. The ride was dangerous and occupied by the Indians. Legend says that she rode the whole way without stopping to sleep or rest, on reaching Fort Savannah, she turned around and with the supplies and an extra horse, started straight back.

Bailey returned a hero and was rewarded with whiskey and the horse she had rode. As a result of her cunning and audacity the Shawnee Indians believed that she was possessed and could not be harmed by a bullet or arrow. After years living on her own Bailey met and married John Bailey in 1785, a "Ranger", and one of the most legendary groups of frontier scouts. In 1788 John Bailey began duty at Fort Clendenin where there was more conflict between the settlers and Native Americans. Bailey was well known for riding up and down the border encouraging men to volunteer their services to join the militia in order to keep the women and children safe. She rode between Fort Savannah and Fort Randolph a distance of 160 miles carrying messages back and forth and knew the paths and trails better than any of the region and was very well respected by the locals. It was during one of these rides that she was discovered and while being pursued by Indians, was cornered, left her horse and found a hollowed out log to hide in. Although the Indians searched for her, she was never discovered, they stole her horse and left. That night she tracked them down and while they were sleeping retrieved her horse and rode off into the night.

John Bailey died in 1802, after which she gave up her home and lived in the wilderness for over 20 years living off the land, one of her favourite places was a cave near 13 mile creek. A local reporter Anne Royall, interviewed Bailey in 1823. During the interview she is reported as saying, "I always carried an axe and auger, and I could chop as well as any man. I trusted in the Almighty. I knew I could only be killed once, and I had to die sometime". She died on November 22, 1825 and was buried in Gallia County. However, later her remains were moved to Point Pleasant. Source, National Women's History Museum (2013).

These women warriors are not the only ones that sent a cold chill down the spine of their male counterparts, history has seen women from various cultures and social environments such as Joan of Arc, The Tru'ng Sisters, Tomoe Gozen, Queen Zenobia and Martha Jane Canary, a.k.a. Calamity Jane are a few that have legends surrounding their names.

Warrior Tribes

Warrior tribes are traditionally egalitarian with no political hierarchy and no social pecking order; individuals as a result of a coup or a personal challenge form them where they have forcibly taken control of several groups that are organised together. In contrast, a state has a political hierarchy with a subordinate political group and power is transferred or taken by coercive strategies, any power obtained by a political group usually has a limited period of time that it can be held for, unlike a dictatorship that has no set time limit. A tribal leader that takes control usually has to fight or use deception to unseat the current ruler. In general, conflict between tribes or states are conducted for one very basic reason "predation" The act of predation involves plundering and marauding known today as 'war' in order to hunt, kill or gain resources, that in turn gave the victor an evolutionary advantage in being able to spread their genes into a wider population.

To some individuals today, the idea of war in any form is repugnant. However, evolutionary psychologists would argue that war is a necessary process to single out and eradicate weak genes in favour of the strongest gene. Warrior traits like those within the women warriors of Dahomey discussed above, would have been passed from generation to generation and in doing so ensured the survival of the gene.

In his book The Selfish Gene, Dawkins (1976) clearly creates a difference between the gene and the human organism, his main theme is that the organisms are designed by genes for the sole purpose of enabling the genes to reproduce themselves. By creating this divide, between the human organism and the gene, he allows for an understanding that the gene is the dominant part of this relationship and has a ruthless selfishness that underpins the physical and psychological processes of homeo-sapiens and to this end, war is a clear and definitive way of ensuring natural selection.

Humans are, along with all animals on the planet, survival machines and when one individual comes upon another in competition over resources, one may well hit back. An idea to keep in mind while considering if human behaviour or some mental trait is an evolutionary adaptation that is being driven by genes and evolution, is a simple question, is it in the genetic interest of the human organism? For example, is it in the genetic interest of an individual to band together with another individual, creating a tribe, to make war on another tribe of individuals? it matters not if the tribes are the aggressors intending to expand their territory and resources or protect their resources.

On the question of rape, is it again in the genetic interest of a man to rape a women and spread his genetic blueprint? I am not for a second supporting this behaviour, however there is a clear difference between what we now know as evolutional behaviour and a moral code by which the majority of humans live. There is an argument that has been put forward by evolutionary psychologist that this behaviour supports human existence.

Local groups banding together make up tribes. In the past, any small group that came together for the purposes of warfare would have been classified as a tribe and villages, settlements or large families could all have created this type of unit. A region that was being threatened by an aggressive tribe would have had good reason to form a tribe, based on mutual associations for the benefit of everyone. They would have been better aligned to protect their women, children, livestock, buildings, homes and farming produce, all of which supports their survival and reproduction. In some cases, a tribe may have consisted of a very large village that had a lot to protect. Tribes were a more effective way for a large population to be successful in warfare, with warriors within the group being escalated to high levels of status. The status of a warrior within a tribe gave them more access to resources, which would include women, food and shelter.

A warrior, although genetically predisposed for violence, would not engage another warrior just for the sake of it. As humans living within a social environment, all individuals have the capacity for violence. They have evolved in the same manner and warriors are not, as common belief would have it, blood thirsty or have a death wish. They do not go around indiscriminately attacking members of their own species for glory. How would any survival machine know that the survival machine that they are attacking is not as strong and mentally equipped as they are.

They have the same chances as any other individual, they may have the same weapons and be skilled in their use. This potential likelihood of injury or death by randomly attacking another member of your own species is a very strong natural selective process, which predisposes an individual to be careful and weigh heavily on the thought of combat and if the potential benefits are worth the risk and outweigh the expected cost. As a species, humans are among the most intelligent to walk the earth and therefore have the capacity to consider if their genetic inheritance will be enhanced by the use of violence, with the warrior who is at the sharp end of the stick taking all the risk.

In a remarkable book by Hobbs The Leviathan (1660) in the chapter "Of the natural condition of mankind as concerning their felicity and misery' he talks about men being equal in faculties of body and mind, that on occasion some men can be stronger and quicker in mind. However, in general when taken together any man can claim what another has, the weakest of men has strength enough to kill a stronger man, this can be achieved by deception and entrapment or by association with others that may also be at risk of threat from the stronger man or tribe.

Hobbs goes on to also consider the strength of mind, which arguably is the more potent of traits when it comes to domination and war stating "I find yet a greater equality amongst men than that of strength. For prudence is but experience, which equal time equally bestows on all men in those things they equally apply themselves unto. That which may perhaps make such equality incredible is but a vain conceit of one's own wisdom, which almost all men think they have in a greater degree than the vulgar; that is, than all men but themselves, and a few others, whom by fame, or for concurring with themselves, they approve. For such is the nature of men that howsoever they may acknowledge many others to be more witty, or more eloquent or more learned, yet they will hardly believe there be many so wise as themselves; for they see their own wit at hand, and other men's at a distance. But this proveth rather that men are in that point equal, than unequal. For there is not ordinarily a greater sign of the equal distribution of anything than that every man is contented with his share" Men therefore seem to have a trait that allows for violence and war, one which is inherited and shows itself when two men or an opposing tribe want the same thing, when this situation arises they become mortal enemies, locking onto a path that eventually leads to either one destroying or subduing another.

Tribes are the vehicles that allow men to obtain dominance and resources over other men and within tribes warriors arise; they step up to the challenge and grow in stature and character. Hobbs goes on to identify three causes of conflict between men and the effects of such a cause "So that in the nature of man, we find three principal causes of quarrel. First, competition; secondly, diffidence; thirdly, glory. The first maketh men invade for gain; the second, for safety; and the third, for reputation. The first use violence, to make themselves masters of other men's persons, wives, children, and cattle; the second, to defend them; the third, for trifles, as a word, a smile, a different opinion, and any other sign of undervalue, either direct in their persons or by reflection in their kindred, their friends, their nation, their profession, or their name" From these words we can gain a sense of the motivations, traits and character that define a warrior and the reasons that drive them into action, for competition, diffidence and glory.

Competition between humans has always been a part of the genetic blueprint and is also found in the majority of other animal species. Men compete for the right to take a women, to own land which is better than the next man, to obtain food, water or fuel from the earth, all of which enables the individual that owns these to out-produce his competitors, produce more healthy children and pass these traits onto future generations. Diffidence is not so well known, why would a warrior pit himself against something that he has a fear of or is unfamiliar with? According to Hobbs it's safety! Defending one's own family or others that help support your family against an unknown assailant is as natural as going to war to increase your resources, the survival machine has this defense automatically built into its genetic blueprint.

The idea that humans are genetically encoded for violence and war to some, will seem like some science fiction film depicting the invasion of earth by aliens, the logic is plain when viewed from the genes point of view, having only one aim for millions of years, replicate, replicate, replicate, at all cost replicate.

The behaviour of humans throughout history has supported the action of war and the development of the warrior spirit, it can be found in the architecture of our structures, the development of our technology, the words and language used to communicate with. One of the great wonders of the world today is the Great Wall of China, built to protect a people from invading warriors with no aim other than to conquer and dominate the lands and the people therein. Although this structure is vast, it is no different from the forts and castles of old or the doors we lock when sleeping for the night, left over behaviours from our ancient past.

Today it's not the tribal band or warring village that invade our fears, although terrorism carried out by a few fanatical individuals has created an indulgence in the act of protection and the lengths that some will go to protect their borders in order to feel safe. No, today it's states and countries, especially those ran by individual dictators who seem bent on gaining as much power as they can that we fear, what has been done to protect us from these countries? Humans have used their intelligence to develop technologies that can build weapons of mass destruction, satellites that orbit the earth to spy on their neighbours, a far cry away from the days of our past, but still this behaviour is imbedded in the way mankind has evolved and the mechanism that helped drive this evolution.

The ancient past of humans is a far cry away from where we are now, back then, nature had set in motion a behaviour that was to forever mould the future of mankind. In our ancient history men were wired for aggression and violence and to all intent and purposes were living in a perpetual state of anarchy, it is from this historical majeure that warriors were born.

Becoming a Warrior

War has been the ultimate mechanism in which a warrior learnt their path, the journey and the methods used to create warriors differed depending on the culture. The Maasai people of Eastern Africa are a Nilotic group that migrated from the Nile region, they are pastoralists, which is a social and economic system based on the herding and trading of livestock. Great value is placed on cattle that are used as a currency to settle most issues that arise in the community. They are also well known for their warrior men, who are raised with the sole intention of becoming a warrior. They live for the majority of their lives outside the main tribe and are not permitted to marry until they are older and have become an elder of the group.

From an early age life as a Maasai boy is focused on becoming a warrior, almost as soon as they can walk they are sent out as herders with calves and lambs. Soon after that ritual beatings begin which are designed to create courage within the young boy. The main right of passage from a boy to a young warrior is achieved through circumcision, which is carried out without any drugs to control pain or the state of the mind.

In his book, My Circumcision, Saitoti (1986) talks about the ritual ceremony of his own circumcision with instructions from his father, "You must not budge; don't move a muscle or even blink. You can face only one direction until the operation is completed. The slightest movement on your part will mean you are a coward, incompetent and unworthy to be a Maasai man. Ours has always been a proud family, and we would like to keep it that way. We will not tolerate unnecessary embarrassment, so you had better be ready. If you are not, tell us now so that we will not proceed. Imagine yourself alone remaining uncircumcised like the water youth [white people]. I hear they are not circumcised. Such a thing is not known in Maasailand; therefore, circumcision will have to take place even if it means holding you down until it is completed."

At such a young age, the mind was being influenced by the powerful use of language, coward, incompetent, unworthy, embarrassment, they also created visual connections, 'imagine yourself'. What great pressure was laid upon such young men. By inference, every boy is being educated in what it takes to be a warrior, which is the opposite of everything they are told. His father then continues to advise him
"The pain you will feel is symbolic. There is a deeper meaning in all this. Circumcision means a break between childhood and adulthood. For the first time in your life, you are regarded as a grown-up, a complete man or woman. You will be expected to give and not just to receive. To protect the family always, not just to be protected yourself. And your wise judgment will for the first time be taken into consideration. No family affairs will be discussed without your being consulted. If you are ready for all these responsibilities, tell us now"

Even the elder tasked with the procedure is as solemn as he can be, bringing the knives to be used in the circumcision a few days before, commanding the young boy to protect and sharpen the blades and if anything should happen to them or they were not sharp enough, then he would bring shame on his family and he would suffer the consequences. If the young boy should not act like a man during the circumcision or fail to turn up, then the family would also suffer; the females would be spat upon and even beaten for having such an unworthy son. The herd of cattle belonging to the family still in the compound would be beaten until they stampeded; the slaughtered oxen and honey beer prepared during the month before the ritual would go to waste; the initiate's food would be spat upon and he would have to eat it or else get a severe beating, all this as a result of not becoming a warrior or not acting like one.

It is no wonder that these people place such a huge responsibility onto their young warriors. The events of Saitoti are best told in his own words "The closer it came to the hour of truth, the more I was hated, particularly by those closest to me. I was deeply troubled by the withdrawal of all the support I needed. My annoyance turned into anger and resolve. I decided not to budge or blink, even if I were to see my intestines flowing before me. My resolve was hardened when newly circumcised warriors came to sing for me. Their songs were utterly insulting, intended to annoy me further. They tucked their wax arrows under my crotch and rubbed them on my nose. They repeatedly called me names. By the end of the singing, I was fuming. Crying would have meant I was a coward. After midnight they left me alone and I went into the house and tried to sleep but could not. I was exhausted and numb but remained awake all night. At dawn I was summoned once again by the newly circumcised warriors. They piled more and more insults on me. They sang their weird songs with even more vigor and excitement than before. The songs praised warrior-hood and encouraged one to achieve it at all costs. The songs continued until the sun shone on the cattle horns clearly. I was summoned to the main cattle gate, in my hand a ritual cowhide from a cow that had been properly slaughtered during my naming ceremony. I went past Loiyan (his sister), who was milking a cow, and she muttered something. She was shaking all over. There was so much tension that people could hardly breathe. I laid the hide down and a boy was ordered to pour ice-cold water, known as engare entolu (ax water), over my head. It dripped all over my naked body and I shook furiously. In a matter of seconds I was summoned to sit down. A large crowd of boys and men formed a semicircle in front of me; women are not allowed to watch male circumcision and vice-versa. That was the last thing I saw clearly. As soon as I sat down, the circumciser appeared, his knives at the ready. He spread my legs and said, "One cut," a pronouncement necessary to prevent an initiate from claiming that he had been taken by surprise. He splashed a white liquid, a ceremonial paint called enturoto, across my face. Almost immediately I felt a spark of pain under my belly as the knife cut through my penis' foreskin. I happened to choose to look in the direction of the operation. I continued to observe the circumciser's fingers working mechanically. The pain became numbness and my lower body felt heavy, as if I were weighed down by a heavy burden. After fifteen minutes or so, a man who had been supporting from behind pointed at something, as if to assist the circumciser. I came to learn later that the circumciser's eyesight had been failing him and that my brothers had been mad at him because the operation had taken longer than was usually necessary. All the same, I remained pinned down until the operation was over. I heard a call for milk to wash the knives, which signaled the end, and soon the ceremony was over. With words of praise, I was told to wake up, but I remained seated. I waited for the customary presents in appreciation of my bravery. My father gave me a cow and so did my brother Lillia. The man who had supported my back and my brother in law gave me a heifer. In all I had eight animals given to me. I was carried inside the house to my own bed to recuperate as activities intensified to celebrate my bravery. I laid on my own bed and bled profusely. The blood must be retained within the bed, for according to Maasai tradition, it must not spill to the ground. I was drenched in my own blood. I stopped bleeding after about half an hour but soon was in intolerable pain. I was supposed to squeeze my organ and force blood to flow out of the wound, but no one had told me, so the blood coagulated and caused unbearable pain. The circumciser was brought to my aid and showed me what to do and soon the pain subsided" after such an experience these young warriors pass into the adult world with all that is expected of them clearly marked before them.

In their past history, before the advent of modern laws and wildlife protection, the next stage in this transference into a fully fledged warrior would have been to bloody their spears by taking the life of a lion, a ritual that many still see to this day as their right. The battle-hardened warriors had a reputation for being fearless when entering battle, prepared to die for their honour.

An encounter with the Maasai warriors given in a book by Huxley (1985) titled, "Out of the Midday Sun", who had moved back to Africa in 1933 after she following an 8-year absence, recalled how she stayed with a district officer called Clarence Buxton, who was considered by the Maasai and some of the locals to be a little mad, he thought that he could change human nature, especially those of African decent.

Clarence had persuaded the local warriors to help him build roads and it was during this time that he had managed to upset some of the warriors. These were young men who had recently passed through the Maasai ritual of circumcision; their need to fulfill the bloodletting and pass into manhood was high. One group challenged another to defy authority (Clarence) and attack him. It did not take long for 40-50 young warriors to amass with their spears and shields ready to take life. Clarence had police fire rifle shots into the air above their heads, she then says " I don't know if he could see the whites of their eyes, but he did see no trace of "homo sapiens", no light of sense or reason, as he had written in a report on the incident, but an expression of 'demonical insensate savagery.

Traditional warrior societies have different initiations, which enable young boys to transfer through and into manhood and eventually become a warrior. Some like the Japanese Samurai, require hereditary through family, for the peasant obtaining such high recognition and honour is not possible. The general theme that runs through this transition is the undertaking of hardship and pain. No society on earth started this process faster than those legendary warriors from Greek mythology "Spartans". Sparta had an early form of ritual that surpassed every other; the whole social formation was based around a military culture. Shortly after birth male babies were bathed in wine to test if they were strong, those that survived this initiation were taken to see the Gerousia, a council of warriors, who decided if the baby would be chosen to be reared as a Sparta warrior. Those deemed weak or who did not conform to what they required, were cast into a chasm on Mount Taygetos.

At the age of seven the boys would enter Agoge system, formally starting their military training. This system was designed with one aim, to produce fearless and skilled warriors, this was achieved by a strict code of discipline and physical toughness. Everything was controlled, they lived in communal groups, were fed a strict diet, undertook weapons training, as well as more intellectual studies like reading, writing and music.

At twenty they joined small groups called the Syssitia, comprising of fifteen other fellow warriors. Within these groups they learnt how to bond and trust each other for support and their life if necessary. Every Spartan warrior had to serve in the military up to the age of thirty, when they could then run for public office and if their life was an honourable one, they could then live with their wives, as up to this time they spent all of their time with other warriors. Only those born as native Spartans were eligible to undertake the training that was laid down by the law within Sparta.

Both the fictional fantasy film and the 1998 comic series by Frank Miller and Lyn Varley portrays this honed military vehicle and warrior spirit when the 300 Spartan warriors engage an army of 300,000 in the Battle of Thermopylae.

Although these are both fictional story adaptations of what is ultimately Greek mythology, there are elements that are correct according to Paul Cartledge a Professor of Greek History at Cambridge University "artistically, 300 is quite powerful, but some of the content is problematic. The movie doesn't really make it clear that although, yes, there were 301 Spartans (300 plus the king), behind those soldiers were about 7,000 other Greeks allied to Sparta. It's also impossible to know exactly what the Battle of Thermopylae was like, but we do know it would have been a very untypical Greek battle because of the terrain: there was a narrow passage next to the sea, only wide enough for two chariots. For two days Xerxes, the king of Persia, hurled his best troops at the Spartans, but because the front was so narrow only a few could reach the fighting" the Guardian, (2007).

What the film does capture is the martial ethos of ancient Sparta and the indoctrinated culture of the Spartan warrior and the military tactics that were employed by a very select and specialised combatant, whose whole early life was dedicated to one end, to become the best he could be, to honour his fellow combatants and fight with absolute commitment to his fellow warriors.

The Language of the Warrior

Before I discuss two warrior cultures, let's take a brief look at the language that surrounds and personifies the image of a warrior and where such words originate. The word warrior originated circa 1300 from an Old Northern French word "werreieor" and in its simplest form means, one who wages war, taken from "warreier guerreir" to wage war.

Consider this text from Native Words – Native Warriors (2007) "the Code Talkers' role in war required intelligence and bravery. They developed and memorized a special code. They endured some of the most dangerous battles and remained calm under fire. They served proudly, with honour and distinction. Their actions proved critical in several important campaigns, and they are credited with saving thousands of American and allies' lives". The passage talks about the Native Indians, the "Navajo" employed by the United States government during World War II as code talkers. Their very unique skill was a specialized form of communications in their own language. Their job was to send coded messages, relaying enemy positions and troop movement; quite often this was being performed under fire. The words used to describe these men are really no different than any other soldier under the same conditions might expect to hear, bravery, proud, honour and distinction. Add to this, valour, magnanimity, courage and we begin to build up an understanding of the virtues and language that come together to create a warrior.

Bravery

Adjective, ready to face and endure danger or pain; showing courage. If bravery was defined as being ready to face danger and pain, then the Maasai warrior during the ritual of circumcision certainly did this. But equally any individual that had a difficult situation to endure could have the label 'brave' attached to them, for instance a women preparing for child birth could be called brave. It's therefore not easy to define a warrior with one word, putting them into a category such as brave. It is more likely the case that a warrior encapsulates many different descriptions of character traits, along with the physical path that has to be trodden as well.

Honour

To have honour is to have the quality of knowing and doing what is morally right, morality can be interpreted differently depending upon cultural values. The word "honour" is synonymous with integrity, honesty, having high principles, ethics, justice, trustworthiness, reliability, dependable and having scruples. In the military, honour is seen as a tangible asset that is rewarded when portrayed by personnel. There are honour codes that are very strict and stringent requiring absolute adherence to the rules and principles of the code.

These codes are used as strict guidelines by which military students and seasoned veterans alike live their lives. Wikipedia (2013) "the honor concept and honor treatise are parts of the United States Navel Academy's honor program. Similar to the Cadet Honor Codes of the United States Military Academy and the United States Air Force Academy, the concept formalizes the requirement for midshipman to demonstrate integrity while refusing to lie, cheat or steel" it is the case that there is an underlying commitment to the words and actions of the code. The honour concept provides service personnel the option of confronting someone who has committed a breach of the honour code, without having to go through the channels of authority and reporting the violation. The penalties for breaking the code can also be quite severe.

The following are extracts from the code;

Midshipmen are persons of integrity: We stand for that which is right.

We tell the truth and ensure that the full truth is known. We do not lie.

We embrace fairness in all actions.

We ensure that work submitted as our own is our own, and that assistance received from any source is authorized and properly documented.

We do not cheat.

We respect the property of others and ensure that others are able to benefit from the use of their own property.

We do not steal.

The military are not the only organizations that have such codes, they have been present in educational societies for centuries, in private clubs and organizations such as the masons, however none of these have to engage in deadly force encounters as part of their everyday occupation.

Magnanimity

Yet another word with Old English, French and Latin roots, meaning being great of heart and mind, definitely a requirement for any warrior. It encapsulates the behaviour that refuses to be petty, willingness to again face danger and for the individual's actions to be at all times noble. The Greek interpretation of the word is "greatness of soul" and according to some historians Aristotle believed that to be magnanimous was the highest of all virtues, it is one that holds above everything the ability to never be drawn into arguments and to have and use the wisdom of Solomon without fear or retribution.

Valour

Yet another word that comes from Middle English which originated through Anglo-French via Latin, with the meaning, worth, worthiness, bravery, to be strong. A particular connotation that goes with valor is strength of mind or spirit that enables a person to encounter danger with attitude and determination, which is often true for military personal involved in war. It may be hard to define spirit in the sense that one can attach it to our understanding of Valor, yet one cannot separate a strong spirit from the sense of what a warrior stands for.

Courage

What is understood by the term "showing courage"? Courage is having the mental and moral strength to face a personal fear, to look danger in the eye! To endure and persevere through hardship overcoming personal difficulties. Courage and bravery go together hand in hand, both bringing to mind an individual that is prepared to walk whatever path is necessary to achieve their goals, to meet head on imminent danger, to walk towards the gunshots and explosions and not away in the other direction. Courage is a Middle English word originally 'corage' from Anglo-French 'curage' first used in the 14th century.

Courage is a state or quality of mind or spirit that enables an individual to face danger, intimidation, pain or fear with self-possession, confidence, and resolution. Usually courage is considered to occur in the face of physical hardships that could result in death or serious injury, however, quite often in conversation we speak of moral courage, having the ability to retain your principles and act correctly in the face of shame, to stand up to those that would put you down. According to Cleary (2009) in his book Training the Samurai Mind, he describes courage as either inward or outward. Outward courage is displayed as "bravery in the exterior appearance and in the face of combat, manifests the force to smash even iron and stone. In addition, one like this is normally vigorous and inclined to forcefulness and has no regard for people. A man with the heart to tear even a wild beast apart.

Of course with the Eastern philosophies the yin also comes with the yang "inner courage is courage not displayed outwardly but kept within the inner heart. Unlike outward courage, this does not make the face stern and solemn and is not an inclination to forcefulness or a manner of speech. It refers to courage "with a gentle face but strong root" clearly courage comes in a variety of guises.

Virtues

The philosophers of ancient Greece considered there to be four cardinal virtues, prudence, justice, temperance and courage, all were required to be maintained by a warrior in the face of adversity and fear. A more up to date investigation into courage has brought forward research into the subject of courage and positive human traits in general with a view to classifying and identifying the traits; courage has been categorised and broken down into four subsections, Bravery, Perseverance, Honesty and Zest.

In 2004 Peterson and Seligman wrote a book, Character Strengths and Virtues in which they introduced the six human virtues that they consider are notable for use in a scientific exploration of positive human traits. They put forward a separation of character categories in order to create clarity in understanding; they make a distinction between Character Strengths, Virtues and Situational Themes. They used a hierarchical classification of positive characteristics based upon the Linnaean system, in which organisms are grouped according to shared characteristics into a hierarchical series of fixed categories, these range from Kingdom all the way down to Subspecies at the bottom.

Virtues are the core characteristics recognised as being consistent in the historical past of humans and are seen regularly in the language of moral philosophers and religious persons. Peterson and Seligman argue that the six virtues of Wisdom, Courage, Humanity, Justice, Temperance and Transcendence are universal across all human cultures, that are grounded in biology through an evolutionary selective process that predisposes these virtues to be selected to ensure that humans are capable of high character that allows them to solve problems that lead to survival of the species. They speculate that these virtues need to be present for an individual to be known as a good character.

Character strengths are psychological processes and mechanisms that enable an individual to display any one of the virtues. Peterson and Seligman state that Wisdom can be achieved by the use of character strengths such as, curiosity, love of learning, creativity, open-minded and a perspective of the big picture of life. An individual that openly displays one of a number of these strengths can be seen as having wisdom and is therefore an individual that could be sought out for advice, they are seen as being virtues.

Situation Themes are the last category in their triad of character traits, these are specific habits that lead individuals to manifest character strengths in given situations. The enumeration of these themes were first studied in the workplace, with a large number being identified by the Gallup organization as very specific to the workplace.

These include 'Empathy' anticipating and meeting the needs of others, 'inclusiveness' making others feel part of the group and 'positivity' seeing the good in others and the group. These could change if the situation were to alter for example, in the context of a family unit. What these themes achieve is a scaffolding of the character strengths that in turn lead to supporting the hierarchical virtues.

What is interesting is the logical process that is involved in habitual traits, that support a strength and lead ultimately to a virtue, the reverse could also be seen as being true for example, negativity as a habit forms a character weakness of desire and depression, that supports one of the seven sins 'Envy' envy being seen as a weakness rather than a strength and one which can lead by desire.

Peterson and Seligman then went on to create what they called the Virtues in Action (VIA) Institute, on visiting their site you are able to take a VIA survey to discover which of the six traits are more in line with your own character. They have also set out a complete breakdown of the six virtues and the corresponding twenty-four character strengths that support the six virtues; their research covers a diverse and multicultural variation of people and societies. Their classifications and the conclusions they have reached deserve repeating here.

1. Wisdom and Knowledge – Cognitive strengths that entail the acquisition and use of knowledge.

   Creativity [originality, ingenuity]: Thinking of novel and productive ways to conceptualize and do things, includes artistic achievement but is not limited to it.

   Curiosity [interest, novelty-seeking, openness to experience]:

Taking an interest in ongoing experience for its own sake, finding subjects and topics fascinating, exploring and discovering.

Judgment [critical thinking]: Thinking things through and examining them from all sides, not jumping to conclusions, being able to change one's mind in light of evidence, weighing all evidence fairly.

Love of Learning: Mastering new skills, topics, and bodies of knowledge, whether on one's own or formally, obviously related to the strength of curiosity but goes beyond it to describe the tendency to add systematically to what one knows; and

Perspective [wisdom]: Being able to provide wise counsel to others, having ways of looking at the world that make sense to oneself and to other people.

2. Courage – Emotional strengths that involve the exercise of will to accomplish goals in the face of opposition, external or internal.

   Bravery [valour]: Not shrinking from threat, challenge, difficulty, or pain, speaking up for what is right even if there is opposition, acting on convictions even if unpopular, includes physical bravery but is not limited to it.

   Perseverance [persistence, industriousness]: Finishing what one starts; persisting in a course of action in spite of obstacles, "getting it out the door", taking pleasure in completing tasks.

   Honesty [authenticity, integrity]: Speaking the truth but more broadly presenting oneself in a genuine way and acting in a sincere way, being without pretense, taking responsibility for one's feelings and actions; and

   Zest [vitality, enthusiasm, vigor, energy]: Approaching life with excitement and energy, not doing things halfway or halfheartedly, living life as an adventure, feeling alive and activated.

3. Humanity - Interpersonal strengths that involve tending and befriending others.

Love: Valuing close relations with others, in particular those in which sharing and caring are reciprocated, being close to people.

Kindness [generosity, nurturance, care, compassion, altruistic love, "niceness"]: Doing favours and good deeds for others, helping them, taking care of them; and

Social Intelligence [emotional intelligence, personal intelligence]: Being aware of the motives and feelings of other people and oneself, knowing what to do to fit into different social situations, knowing what makes other people tick.

4. Justice - Civic strengths that underlie healthy community life.

Teamwork [citizenship, social responsibility, loyalty]: Working well as a member of a group or team, being loyal to the group, doing one's share.

Fairness: Treating all people the same according to notions of fairness and justice, not letting personal feelings bias decisions about others, giving everyone a fair chance; and

Leadership: Encouraging a group of which one is a member to get things done, and at the same time maintaining good relations within the group, organizing group activities and seeing that they happen.

5. Temperance – Strengths that protect against excess.

Forgiveness: Forgiving those who have done wrong, accepting the shortcomings of others, giving people a second chance, not being vengeful.

Humility: Letting one's accomplishments speak for themselves, not regarding oneself as more special than one is.

Prudence: Being careful about one's choices, not taking undue risks, not saying or doing things that might later be regretted; and Self-Regulation [self-control]: Regulating what one feels and does, being disciplined, controlling one's appetites and emotions.

6. Transcendence - Strengths that forge connections to the larger universe and provide meaning.

Appreciation of Beauty and Excellence [awe, wonder, elevation]: Noticing and appreciating beauty, excellence, and/or skilled performance in various domains of life, from nature to art to mathematics to science to everyday experience.

Gratitude: Being aware of and thankful for the good things that happen, taking time to express thanks.

Hope [optimism, future-mindedness, future orientation]: Expecting the best in the future and working to achieve it, believing that a good future is something that can be brought about.

Humor [playfulness]: Liking to laugh and tease; bringing smiles to other people, seeing the light side, making (not necessarily telling) jokes; and

Spirituality [faith, purpose]: Having coherent beliefs about the higher purpose and meaning of the universe, knowing where one fits within the larger scheme, having beliefs about the meaning of life that shape conduct and provide comfort.

Different cultures around the world have certain ideas of what it means to be a warrior, from the above research the virtues that are all too evident in the warriors of today are Courage, Temperance, Wisdom and Knowledge, Justice and Humanity, it could also be argued that a few of the character strengths in Trancendence are also found in the traits of a warrior. Certainly a warrior has many of the above virtues and strengths of character that cultures around the World would agree should be ubiquitous in most warriors, especially those that are charged with defending the general population and having to enter the theatre of war. This makes the traits of a warrior universal across cultures and part of human knowledge surrounding the mindset and character of this type of individual.

The ideas of what virtues are have changed over the years according to which philosopher is discussing them. Back in 350 BC Aristotle defines virtue as the middle ground, or "golden mean" the place between two extremes, courage being half way in between fearfulness and rashness. Circa first century, an anonymous Hellenistic Jew writes 4 Maccabees, an apocryphal book of the bible. It reads "Now the kinds of wisdom are right judgment, justice, courage and self-control. Right judgment is supreme over all of these since by means of it reason rules over the emotions" Jacobson (2013). Being able to apply the right judgment in times of high stress may not in fact lead to courage, as one could argue that some situations should be avoided at all cost, rather than jumping in head first thinking that you are being courageous.

Virtues for warriors first make an appearance between the 13th – 16th centuries within the Samurai code of medieval Japan, according to Bushido, a warrior should act with calmness, fairness, justice and propriety at all times, it also encourages learning of the artistic, music and literary arts as a symbiotic part of a warrior's discipline. Bushido means "the way of the warrior" and represents a Japanese culture, which encapsulated a Samurai's life. Bushido is a relatively new term and was the code by which the Samurai lived their life.

This code grew over many centuries and was influenced during this growth by Taoism, Confucianism, Buddhism and Shinto, together with military tactics and martial arts. In ancient times, the warrior traditions of the samurai set them apart from other warriors, partly due to their willingness to embrace death. Budo was the first term used to describe the Samurai (warrior's way) later came Shido (knight's way), which eventually cumulated, into Bushido. The word Samurai originally meant "attendant" Cleary (2009) referred to the armed retinues of aristocrats. The writings of ancient Japan and the samurai caste are huge and allow for a rich understanding of the culture of that time, however I want to convey a sense of who the Samurai were, identifying their warrior culture and thought processes and to do this best I will quote some of the writing surrounding the Samurai.

Shiba Yoshimasa (1349- 1410)

To regard your one and only life as like dust or ashes and die when you shouldn't is to acquire a worthless reputation. For a person to be bad tempered is more disgraceful than anything. No matter how irritated you may be, at the very first thought you should calm your mind and distinguish right from wrong. If you are in the right, then you may get angry.

Ichijo Kaneyoshi (1402 – 1481)

Honesty just means a straightforward mind. If the mind is distorted, all behaviour is distorted.

Nakae Toju (1608 – 1648)

Culture without warrior hood is not true culture; warrior hood without culture is not warrior hood.

Warrior hood that is contrary to justice may be called warrior hood in name but it is not warrior hood in reality.

To say that there are a lot of people with hidden courage means that you cannot tell just by a superficial view of appearances. There are surly cowards among those who appear easygoing, while there will also be courageous ones. There are also bound to be cowards among those who appear to be fierce, as well as those who are indeed mettlesome. The perceptivity to aim for is the ability to see whether someone is courageous or cowardly at heart.

Because the humaneness of their enlightened character is clear, this courage naturally exists within humanity and justice, so it is called the courage of humanity and justice. Because it is such supreme courage that it has no rival in the world, it is also called great courage. True warrior hood is nothing but this, great courage.

In the book of the Shadow of the White Flag it says that a skillful attack cannot match warlike soldiers, warlike soldiers cannot beat elite knights, elite knights cannot match a disciplined system, a disciplined system cannot oppose humanity and justice. It is for this reason that Sun Tzu's Five Things, the Way is first, while in Wu Tzu's Art of War, Harmony is first, whether called the Way or Harmony, in either case this refers to the virtues of humanity and justice.

Kumazawa Banzan (1619 – 1691)

A good warrior is always courageous and deeply devoted to the way of the warrior and martial arts; he takes care not to stumble what ever happens, respects his ruler, pities everyone from his wife and children to the old and young all over the world, and he prefers peace in the world from a humane and loving heart.

They say hidden courage is the best courage, but even a sword with a fine blade may be found to fail when tested. It seems that the relative strength of people's martial courage is also like this.

These are writings from authors covering a few centuries of Japanese history, we then have the 'Hagakure' a book written by Yamamoto Tsunetomo, translated by Wilson (2002) According to Wilson Yamamoto Tsunetomo had entered service to Nabesshima Mitsushige, the then third Daimyo of the area know as Saga Prefecture. Tunetomo was 42 when the Daimyo died. He was prohibited from committing ritual disembowelment upon his master's death and chose to leave the Samurai caste and become a Buddhist priest.

While in semi-seclusion another young Samurai, who was a scribe, came to live with him, their conversation lasted seven years after which time the scribe wrote the book based on their conversations and gave it the title 'Hagakure' which could mean either 'hidden by the leaves" or "hidden leaves". Even from this short introduction we can obtain a sense of the warrior spirit that pervaded all during these ancient times.

Yamamoto Tsunetomo, translated by Wilson (2002)

The way of the Samurai is found in death.

If by setting one's heart right every morning and evening, one is able to live as though his body were already dead, he gains freedom in the way.

When one's attitude on courage is fixed in his heart, and when his resolution is devoid of doubt, then when the time comes he will of necessity be able to choose the right move.

Even if one's head were to suddenly be cut off, he should be able to do one more action with certainty.

The heart of a virtuous person has settled down and he doe's not rush about things.

Step from under the eaves and you're a dead man. Leave the gate and the enemy is waiting. This is not a matter of being careful, it is to consider oneself dead beforehand.

Indians

There are so many warrior classes throughout history that it's difficult to settle on just a few, the intention of this chapter was to convey a sense of what it should be like to be regarded as a warrior and to provide an understanding of some of the strengths of character that in times of old separated the warrior class from the general population. Before we move on let's take a final look at one more class of warrior the American Indian. For thousands of years, American Indian men have protected their communities and lands. "Warrior" is an English word that has come to describe them. However, their traditional roles involved more than fighting enemies. They cared for people and helped in many ways, in any time of difficulty. They would do anything to help their people survive, including laying down their own lives.

Among the Indians of the North American plains in the eighteenth and nineteenth centuries, a boy was taught from infancy that bravery provided the route to social success among the tribe, old age was evil and was to be avoided and that it was much better to die young, preferably on the battlefield defending the honour of his tribe. They were taught to admire and emulate great warriors and that they should follow in their footsteps. Warriors were regarded with the utmost respect in their communities.

Boys trained to develop the spiritual, mental, emotional, and physical strength they would need to become warriors. Many tribes had special warrior societies, which had their own ceremonies, songs, dances, and regalia. Usually, a warrior had to prove himself before being asked to join a warrior society. It was a great honour to be chosen in this way. Any boys that were of lesser character were identified and sent off to help with menial tasks. In those ancient times, one tradition of passing from boy to warrior was to complete four tasks, on achieving these he would then become a warrior chief, they were to touch an enemy while alive, to disarm an enemy, to lead a successful war party and to steal a horse.

Native American Indians had a wide and diversified culture that was specific to different tribes. However most had some type of association with nature and included ritual dances such as the Sun Dance, War Dance or Ghost Dance and of course let's not forget the Rain and Sun Dance, traditions surrounding Eagle Feathers, Medicine Men, Vision Quest, Sweat Lodges, Peace Pipes, Animism, Dream Catchers, the list goes on. What is important is that the whole tribe worked as one with the hierarchical warrior system heading the people. To explore briefly the spirit and ethos of the Indian warrior I want to look at one particular individual, Medicine Crow.

Medicine Crow was the last war chief of the Crow tribe, this in itself is no ordinary title given to a brave that has experienced war. To obtain this title an Indian has to achieve four tasks and live to tell the tale. Medicine Crow was born in Montana in 1913 and was raised in the warrior traditions of the Crow. According to Indian Country (2014), his grandfather was Medicine Crow, a renowned fierce warrior and scout during the Plains and Indian wars in the 19th Century. "My grandfather trained me to be a warrior," notes Joe Medicine Crow. "The Crow people were so-called, 'warlike.' We were a very militaristic people." During childhood he underwent strenuous training that was not unlike the methods employed by the ancient Spartan culture, he was put through arduous feats of strength, pile-driving buffalo, running barefoot over long distances, swimming dangerous rivers, riding bareback, all designed to build his spirit and the virtues of a warrior within the tribe of the plains Indian. He was taught to acknowledge and control fear, when faced with imminent danger, to hunt and to survive harsh environments, to train his mind and body.

Just before the outbreak of World War II he was studying for a degree before volunteering to join the army. He was sent to Europe and was soon in the thick of the fighting on the front and it was here that he completed the four tasks to become a war chief. The four tasks are, to lead a war party against the enemy and succeed, steal a horse belonging to the enemy, disarm an enemy and touch an enemy without killing him, this is known as 'counted coup' winning of prestige. All these were achieved during his time on the front in Europe. Whenever he went into battle, he wore his war paint beneath his uniform and a sacred eagle feather beneath his helmet.

Indian Country (2014), during an interview with Medicine Crow described how the counted coup was achieved. Medicine Crow saw a lone German soldier walking past in a narrow alley as he hid waiting to ambush someone. "I saw his rifle and I knocked it out of his hands," he recounts. "All I had to do was pull the trigger, but for some reason I put my gun down and tore into him." After a violent struggle, Medicine Crow held the German soldier's throat by his hands, ready to finish him off. The soldier gasped, "Momma!" and Medicine Crow let him go out of sympathy. With that deed and without meaning to, he had committed two of 4 deeds to becoming a war chief.

What is evident from the life of a warrior is that training and mindset leads a person throughout their life, virtues and character emerge from the balance achieved when the journey includes both the martial and spiritual path. Being able to engage a warrior's mindset can help give direction to many choices in normal everyday life as well as being capable of entering this mindset when the situation requires it. Being adaptable and instinctively knowing when to enter this state of mind is what the warriors know within themselves and is a byproduct of the path itself.

6 VIOLENCE IT'S NATURAL LET IT BE

"If only it were all so simple! If only there were evil people somewhere insidiously committing evil deeds, and it were necessary only to separate them from the rest of us and destroy them. But the line dividing good and evil cuts through the heart of every human being. And who is willing to destroy a piece of his own heart?" Aleksandr Solzhenitsyn, The Gulag Archipelago: 1918-1956.

Having spoken about the path and life of a warrior, it's relevant that I include a chapter on violence and aggression, I do not want this chapter to dissolve into a story of evil and the good fight to protect the innocent, however it is important that we at least have a fleeting glance into the dark side of human nature. Ultimately this book is about the ability for humans to fight and behave in a manner that confronts violence and the science involved in training and teaching others to have that same ability. It's also important that we understand what we are training for, as violence covers a huge and diverse spectrum of human behaviour, just like our own perspectives are different given the experiences that shape them, violence is different depending on the culture and the mechanisms that drive it.

To some, any violence is wrong, any aggression is wrong, evil has to be banished from our lives forever, they sit firmly on the fence that believes God sees all and that one should not do harm. This is certainly not going to be a chapter on religion and violence; instead I will focus a little more on the evolutionary explanations for violence within the species homeo-sapian.

During the previous chapter I introduced Hobbs to the reader and his short statement on the nature of man, this helped to set the scene on why man undertakes violence and the reasons for such actions. Violence in itself is a vast subject and one that has had many books dedicated to the subject, what sticks in my mind is a saying that was first introduced by Jainism, a religion traditionally known as Jaina dharma, originally from India. Jainism teaches that the path one should follow is one of non-violence and peace towards all living things, the story is about a king who was once trying to understand how people saw things in the world. He invited five blind men to his palace, he then asked each one to touch an elephant and then to describe what they felt based on their experience. The first touched the trunk and said "elephants are like snakes" the second one who grabbed the tail said "no, to me it feels more like a rope" the third felt the side of the elephant and said "it feels like a wall" the forth put his arms around the leg and said "elephants are more like pillars" the last felt the ear and said "elephants are more like winnowing fans" Hardy (2011). This analogy was also used by Rory Miller in his book Meditations on Violence, to put over the simple fact that violence means different things to different people and that it is a vast subject.

The first thing that should be done is to categorize violence, this will enable us to approach the subject from a specific viewpoint. There is no doubt that many may disagree with this, however we have to start somewhere. Violence has two very basic categories that can then be broken down into further subcategories, a project in the making is the categorisation and explanation of these, so I will not dwell too long here. One area that we will need to cover is why? Why do humans kill? Not only do we kill other humans we also kill in some way most other members of the species 'Animalia' all creatures great and small, as well as the plant and ecosystems that supports all life.

Category 1 Social Violence.

Social Violence (SV) encompasses all the violence that as a social people we have come to accept as part of the behaviour of humans. In this category we find all types of sport that involves human aggression, boxing and MMA are the most popular due to their media exposure. We have violence within the community in which we live, especially when it involves a fight or aggression towards another individual, and although not a comfortable subject, aggression and violence by the young from very early ages until they reach adulthood. There are some cultural differences that will become apparent but in the main that covers a lot of violence.

Category 2 A Social Violence.

A Social Violence (ASV) is everything that is not in the social category. The type of violence that usually provokes a violent reaction is found in this level, here we find War, murder, rape, sexual abuse, killings that involve mutilation, gross, graphical use of blades, genocide, infanticide, what is interesting is that many of the words that we use to describe killing end with 'cide'. This word is again taken from Latin, it is used as a suffix that means "a killer of" or "a person or thing that kills".

Cultural Excuses

As a species, we are undoubtedly violent, however this propensity to violence is within some cultures considered normal, it is only what could be called "civilized society" that considers violence to be unhealthy and immoral. Countries that consider themselves to be civilized are also the very same countries that have, in some cases, the highest amount of recorded crime. The U.S for example has arguably the highest crime rate in the world and although this may have seen a slight drop in some years it still remains high and has started to escalate. People commit all types of violent acts against animals, plant life, property and even themselves, this chapter will only be concerned with violence against other members of the human species and let's not forget that often violence is carried out by more than one individual, it can take the form of organized violence by a group or gang, all the way to violence committed by the state, 'War'

Depending upon how severe we class violence and also include aggression, we may find that harmful behaviour also falls within this category. If we are destroying the environment we are also, by default, killing other humans as a result, drought, pharmaceutical discharge, unsafe food, are we not all contributing in some way to this type of violence?

Violence regularly features as a bi-product of a game, within football in the UK violent fights often continue well beyond the finishing of the game and into the streets and pubs of the local area. Not a night goes by where the local or national news is not reporting on violence. It pervades our homes through the box in the corner of the room invading like a swarm of locus; it seeks out young and old with no care for the consequences, by portraying violence within children's cartoons and entertainment in general, which includes video and computer games. Is it no wonder that our society is still a violent one, having said that according to Gardener (2008) in his book the Science of Fear, we are living in the safest times throughout the whole history of mankind?

Evolution

Darwin introduced the concept of evolution by natural selection in 1858 and from this time man has developed theory upon theory as to what processes have enabled humans to become the dominant species on the planet. One of the supporting arguments is that living beings tend to produce more offspring than the environment can effectively support with its natural resources. The result of competition means that violence has its place in providing a means to resolve conflicts between those that have resources necessary for survival and reproduction and those that don't. This trait for violence would, according to Darwin's theory, select those that have the capacity for violence over and above those that do not, therefore over thousands of years of evolution stone age man became more and more adapt at bringing violence to bear on his competitors in order to survive and have the resources required for reproduction.

In his book Selfish Gene, Dawkins (1976) describes the individual, as a selfish machine programmed to do whatever is best for its genes as a whole. He puts it in very clear language, to a survival machine, another survival machine which is not its own child or another close relative, is part of its environment, like a rock or a river or a lump of food, it is something that gets in the way, or it is something that can be exploited, it differs from a rock or a river in one important respect, it is inclined to hit back, because it too is a survival machine that holds its immortal genes in trust for the future and it to will stop at nothing to preserve them. Natural selection favours genes that control their survival machines in such a way that they make the best use of their environment, this includes making the best use of other survival machines, both of the same and of different species. With this in mind it is no wonder that mankind became so adapt at the tool of violence, as in its simplest form this is exactly what it is, you can also now begin to see how this behaviour begins to show itself in our very young children, this I cover in more depth in Chapter 10 The Bully.

Before I move on, it's important that violence in children is put into perspective as it would seem that this wiring for violence is inherent in every child and begins to show itself very early during the terrible twos. According to Pincker (2012) the psychologist Richard Tremblay has measured rates of violence throughout the normal lifespan of humans, rather concerning is his conclusion that the most violent period in an individual's life is that time when they are two, we may have expected this period to be in adolescence or young adult hood, but no, it was when we were hardly old enough to even talk. The usual behaviour shows itself as hitting, biting, kicking and general moody behaviour, this trend of violent behaviour then begins to slowly decrease throughout the infant's early years. It is therefore of little comfort that these typical children are not capable of wielding any physical tool of violence or else we could see a high number of two year old, on two year old, killings.

The process of natural selection has to also have the ability to pass traits on to future generations, if this were not the case, we would never have adapted to our environment. What this means is that there has to be a mechanism to pass genes on with, this is achieved through heritability and first shows itself during that all too well known stage of the terrible twos.

The blank slate theory was once popular, it was originally believed that parents or any significant caregiver could harm their children by mistreating them, which of course is absolutely true. It was the philosopher John Locke who first proposed the idea that any child could be molded into whatever person he desired, politician, soldier, scholar, this theory became known as the "Educationalists" view and considered that every child born was a "blank slate" and as such they were intellectually and morally clean with no preconceived ideas or knowledge.

This understanding that a child was neither good or bad was at opposite ends to the popular religious views at the time who considered that a child was born inherently bad and that their sins had to be beaten out of them, subduing them to the rod and will power of their masters. A child's nature therefore is easily manipulated during its early years and those that are close to them have the ability to affect intelligence, social skills, mental abilities and personality. This has been discovered to be not true, however due to later studies that researched separated and adopted children, it was found that they mirrored their peers in values and social identities, indicating that social interaction helped develop these children according to the doctrines and culture of their caregivers, not of their parents.

Pincker (2012) states that studies of adopted children showed that they ended up with personalities and iQ scores that are correlated with those of their biological siblings but un-correlated with those of their adopted siblings, this confirms that adult personality and intelligence are more a result of genes than of social environment. What has this got to do with violence? Well there is a theory that violence has a genetic marker and that it is inherited rather than learned.

The survival machine, which is man, has to compete for resources, this includes finding a suitable female to reproduce with and in achieving this will in turn ensure a degree of inheritance of the genes that are found within themselves "the survival and reproductive success (RS) of different heritable variants is not random; some properties systematically enhance their bearers' ability to survive and reproduce" Kurtz (1999). What this tells us is that if a survival machine that lives in an environment where competition for resources is part of their everyday survival, can by the use of the tool of aggression or violence exclude another from access to those resources, including potential mates, then it is likely that those that were unable to fight to protect their resources would not survive, what we see here is that everything has a cost.

Heritability has an effect over many generations on the results of differential reproduction, the selfish organism and its offspring profit by an increase in the success variants and a decrease of eradication over time of the less successful variants.

Krutz goes on to say that evolutionary theory views the evolution of violence in the context of its costs and benefits measured ultimately in terms of RS. It is, however, not about a single individual but about the average success of certain design features. An adaptation for aggression can be advantageous to the survival of copies of the genes that lie on its basis even if some individuals perish because of it. Evolutionary theory is also different from the outdated idea that selection would optimize the good of the species. The 1960s discovery of kin selection, the fact that selection operates on the combined or 'inclusive' fitness of the individual and its relatives, has led to new hypotheses and findings of the patterns of conflict, competition, violence, peace, and altruism.

Taking these findings and applying them to humans indicates that violence is indeed an intrinsic part of our nature and the reason why we have achieved dominance over most life forms that inhabit the planet, we are very effective competitors and have devised ways to ensure that our hold on the planet remains concrete to the point of destruction.

The rewards of violence include control over nearly all resources, where shelter/housing is built, access to mates and the ability to choose our destiny, all this with a potential cost of an increased risk of death or injury as we negotiate our journey with other humans who have the capability to be just as violent. "hence, evolutionary variants that would aggress indiscriminately, without paying heed to the costs of aggression, would be less likely to succeed than those capable of calibrating their responses and aggressing only when benefits are likely to be higher than costs. Thus, we should expect that psychological decision mechanisms would have evolved that are sensitive to the prospects of success and the size of risks involved in different behavioural options. The basic options are often attack (or counterattack), submission, or flight" Kurtz (1999)

Competition has a side effect and that is, it can create fear in both the aggressor and the victim, if one survival machine covets the land of another survival machine, the owner of the land being threatened may feel that his competitor is going to eliminate him by killing him and taking over his resources, he may then start to believe this and make plans to eliminate their perceived aggressor first, in a pre-emptive strike, up until this point never had to defend his resources. The paradox here is that the other survival machine may be thinking exactly the same as you, laying plans for his own pre-emptive strike, even though he would not normally be drawn into any conflict. This counter thought by counter thought can go on to the power of infinity! This paradox is often referred to as the 'Hobbsan Trap' and according to Morey (2000) it could also be termed as a 'security dilemma'

This dilemma is a common one and can be found in every walk of life where violence is a part of the culture, from individuals that face each other off in a test of whose got the largest ego, where one thinks the other is about to start, and the other thinks the same, they continue until they both blow out of steam or a trigger occurs in one, and off they go. Step the violence up to two individuals that are armed with a blade or worse still a gun, each believes that the other is only moments away from launching the attack and the situation is getting worse by the second. This sequence of violence can be taken as far up the scale as a stand off of the two largest super powers in the World, back when Russia was still in the arms race with the US for example.

For thirteen days in October 1962 the world waited, seemingly on the brink of nuclear war and hoped for a peaceful resolution to the Cuban Missile Crisis.

In October 1962, an American U-2 spy plane secretly photographed nuclear missile sites being built by the Soviet Union on the island of Cuba. President Kennedy did not want the Soviet Union and Cuba to know that he had discovered the missiles. He met in secret with his advisors for several days to discuss the problem.

After many long and difficult meetings, Kennedy decided to place a naval blockade, or a ring of ships, around Cuba. The aim of this "quarantine," as he called it, was to prevent the Soviets from bringing in more military supplies. He demanded the removal of the missiles already there and the destruction of the sites. On October 22, President Kennedy spoke to the nation about the crisis in a televised address. No one was sure how Soviet leader Nikita Khrushchev would respond to the naval blockade and U.S. demands. But the leaders of both superpowers recognized the devastating possibility of a nuclear war and publicly agreed to a deal in which the Soviets would dismantle the weapon sites in exchange for a pledge from the United States not to invade Cuba.

In a separate deal, which remained secret for more than twenty-five years, the United States also agreed to remove its nuclear missiles from Turkey. Although the Soviets removed their missiles from Cuba, they escalated the building of their military arsenal; the missile crisis was over, the arms race was not. John F. Kennedy Presidential Library and Museum (2014).

The above is a clear example of the Hobbsan dilemma, the usual policy of deterrents had failed and each side had progressed to the stage where, the other had not carried out a pre-emptive strike, but each felt that they were strong enough to survive a fist strike and then carry out a complete retaliation to a higher level than was threatened against them. The key to deterrents is the credibility of the threat that you will retaliate, if for a moment your enemy thinks that you will not carry out any retaliation, then you leave your competitor with the thought that you are vulnerable and they have no fear of you committing to a pre-emptive strike. Worse still, if your enemy believes that once they have carried out their initial attack you are going to sit back and think that it's no longer worth it to expend more energy in retaliation, then this leads to the greater possibility that an attack will be launched.

The way in which this type of dilemma can be avoided is to create the image and reputation that if attacked, you will not just lie down and take it, in fact you will aggressively react to infringements upon your rights with ruthlessness and aggression. That way your enemy is always kept on the back foot, believing that whatever his response yours will be as violent, if not more so than his. What this strategy implies is that the cost to the aggressor on you would cancel out the benefit gained from their action; the resources would not prove valuable enough to take the risk. The most important understanding that has to be known is that your threat of retaliation is very credible. Indeed, if you have had to retaliate before and have done so effectively, your reputation will proceed you and as a consequence you will be safer and your resources will be protected.

If resources and reproduction are the main goals of an organism, are we the only animals on the planet that have this capacity for violence? Or is there evidence within other species that leads to the conclusion that maybe we are genetically constructed to plan and carry out the tool of violence.

Our Ancestors.

Although our ancestors have become part of our fossil history and are now extinct, they still have a story to tell in both their archeological evidence and their descendants that are still around today, namely chimpanzees and the primate lineage. It is probably the case that homeo-sapiens did not evolve from apes, instead there was a divergence in our early evolution that created two separate lines of descendants, man and ape. Let's now focus on the violence that has been evidenced by research and observations carried out on our ape relatives.

Chimpanzees live in communities just like man and they can number up to 100 or more in a common group, they also occupy rigid and identifiable territories that are patrolled and marked, just like our own borders between countries. This territory contains the resources that are required for the troop to survive, and reproduce. Gathering and foraging also forms a major part of the behaviour of chimpanzees within their wild habitats.

Chimpanzee cultures are therefore much like those of humans and they have a highly developed mental capacity in comparison to other animals. They are often seen to use tools to achieve a goal, such as rocks for cracking hard nuts and sticks as devices to extract ants or honey from hard to reach places. They have mental traits and have been shown to have the ability to reason, show generalization, symbolic representations and have a sense of identity.

Dr. Jane Goodall was the first to study primates extensively in the wild, born April 3 1934 in London she first travelled to East Africa in 1957. Throughout her years of studying chimpanzees she discovered that they displayed a wide range of complex emotions that were once thought to be uniquely human. For instance, chimpanzees clearly exhibit emotions such as joy, sadness, fear and despair. Chimps have also been found to possess an almost human-like enjoyment of physical contact, laughter, and community. These emotions have been evidenced particularly in chimpanzees, more so than other mammals, due to their facial expressions and their facial musculature that is so similar to ours.

Their behaviour involves splitting into groups and exploring the jungle for food, if one group happens to come upon another group from a different community on the border between territories, the interaction between the groups often becomes aggressive and threatening. Pinker (2012) informs us that this type of encounter nearly always ends in a noisy battle between the two groups, especially when their numbers are the same, the two sides bark, hoot, break branches, throw objects and charge at each other for up to half an hour, until eventually the smaller group decides to return to the safety of the undergrowth. These types of displays are found throughout the animal kingdom, he goes on to say that when this type of encounter involves a pair of single males, the violence can escalate and in some circumstances one or both can receive a serious injury or even be killed during the ensuing fight.

It was originally believed that primates did not kill each other and it was Goodall that first observed the true nature of the chimpanzee. She observed that when a larger group encounters a smaller group or a single individual, the behaviour is very different. Pinker goes on to cite her, "no longer do they fall into tit for tat exchanges hooting and charging, instead they take advantage of their superior numbers". If it's a female that is encountered they may try to seduce her and mate, if she has a baby with her they will usually attack both her and her baby, eating the newborn. Worse still, if they come upon a solitary male or isolate one from a small group they will go after it with murderous intent and savagery, two attackers hold down the victim, others will beat him, bite off his toes and genitals, tear flesh from his body, twist his limbs, drink his blood or rip out his trachea.

In one community a dominant group targeted every male in a neighbouring group and killed every last one. If this type of mass killing had occurred among humans, the perpetrators would be labeled evil and the act one of genocide.

In the nature of man there are known to be three causes that initiate humans on their quest for violence.

1 Gain
2 Predatory raids, safety pre-emptive raids and reputation.
3 Retaliatory raids.

During the time before man lived in a world without a controlling power, such as a government, to keep him in line, society was in a perpetual state of existence called "war" and in such condition they lived in continual fear and danger of violent death. Hobbs cited by Pinker (2012)

The above evidence of primate behaviour points clearly in the direction of supporting the point of view that violence is not just a human capacity and that it is inherited and transferable throughout our species. It also seems to have been a part of our behaviour from as far back as we can imagine, and in today's civilised societies this long past propensity for violence does not sit well with some. Over the past century we have seen the rise of all types of groups trying to bend the tool of violence to fit with their personal views, never really grasping the enormity of this issue. Human rights have become the biggest PC of our past history, however before we can get a little more up to date, I believe it will be useful to explore the transition from savage rampaging tribes set in a time of anarchy, to the rule of law governed by the state. Somewhere along this timeline, violence changed, it became more civilized, at least in comparison to the pre-civilised time period as described by Hobbs above.

It is a widely held belief by archeologists that humans lived in a state of anarchy up to approximately five thousand years ago. This has been evidenced by examination of the cause of death on the fossil history of our caveman ancestors. The existence of humans was originally as hunter-gatherers, these tribes of early humans were in existence way before the agricultural revolution and during this time man relied solely on what nature could provide and his survival instincts. Today there are still a few tribes that have survived the civilising process and remain as we all were once upon a time.

The Kalahari bushmen of the Kalahari Desert of Africa use one of the oldest forms of hunting known to man, persistence hunting. Before guns, before knives, before weapons, we used our bodies to hunt. More specifically, we chased our prey until it collapsed. Kalahari hunters chase their prey (typically a kudu i.e. antelope) for 2 to 5 hours over 25 to 35 km in temperatures of about 40 to 42°C (104 to 107°F). Humans actually use much less energy to increase speed than most animals. Their sweat also helps keep the runners cool, whereas the animal must take a rest and find water. The ideal time for the Kalahari to go for the kill, spearing the beast at close range. Persistence hunting is thought to have been one of the earliest forms of human hunting, having evolved 2 million years ago Czartoryski (2011).

What is also interesting here is that evolutionary psychology, a relatively new field, uses persistence hunting as an explanation as to why humans lost their hair, as they needed to sweat to maintain this long distance hunt. A human ancestor that had large amounts of hair covering the whole body would not have been able to hunt effectively, therefore evolution would have favoured the human with less hair.

The Sentinelese dwell on the Andaman Islands in the Bay of Bengal between India and the west of Burma. They are believed to be one of the last Stone Age tribes in the world to maintain their independent existence and are ruthless when it comes to defending their territory. Incredibly, they also managed to survive the 2004 Indian Ocean Tsunami. The Sentinelese are the only hunter-gatherers who resist complete contact with the outside world. In 1974, when a film crew attempted to make friendly contact with the Sentinelese, leaving gifts of food and some pots and pans, one of the islanders shot the film director in the thigh with an arrow. A year later when tourists began nearing the area, they were repelled with Sentinelese arrows. Czartoryski (2011).

Historians have speculated that 20,000 years ago hunter-gatherer tribes consisted mainly of small self contained groups. These were made of small families bound together for the benefit of protection and survival advantages. Gradually, these small tribes began to develop into larger communities with neighbouring tribes that had a similar language or were of interest to each other, one tribe may have had resources that one tribe did not have, access to water for example, or an alliance was necessary due to a competing or warring tribe threatening the existence of a nearby tribe. Inter tribal marriages began to take place along with trade and exploration. Communities gathered pace and population increased to levels where individuals left to seek their own resources, creating extended families and ties to other areas of the land and larger numbers of tribal members that would help defend territory.

These hunter-gatherer tribes did not at first stop long in any one location, they did not cultivate the land and instead gathered what could be found readily available within their territory. Men were responsible for the hunting and women and children the gathering of vegetation, roots and seeds. The time period according to Diamond (1997) was around 11,000 Bc when village life began to appear in certain parts of the World.

The first population of the Americas began at this time, which coincided with the end of the Pleistocene era and the last ice age, it was approximately 2,000 years from this date, that plant and animal domestication began.

Before early humans could expand in large numbers and begin to domesticate animals, they had to overcome the problem of not only feeding themselves but also supplying enough food to rear animals and this meant the domestication of plants first.

Plant Domestication

The wild progenitors of crops including wheat, barley, and peas have been traced to the Near East region. Cereals were grown in Syria as long as 9,000 years ago, while figs were cultivated even earlier; prehistoric seedless fruits discovered in the Jordan Valley suggest fig trees were being planted some 11,300 years ago. Though the transition from wild harvesting was gradual, the switch from a nomadic to a settled way of life is marked by the appearance of early Neolithic villages with homes equipped with grinding stones for processing grain.

The origins of rice and millet farming date to the same Neolithic period in China. The world's oldest known rice paddy fields, discovered in eastern China in 2007, reveal evidence of ancient cultivation techniques such as flood and fire control.

In Mexico, squash cultivation began around 10,000 years ago, but corn (maize) had to wait for natural genetic mutations to be selected in its wild ancestor, Teosinte. While maize-like plants derived from Teosinte appear to have been cultivated at least 9,000 years ago, the first directly dated corncob dates only to around 5,500 years ago. Corn later reached North America, where cultivated sunflowers also started to bloom some 5,000 years ago. This is also when potato growing in the Andes region of South America began. National Geographic Society (2014)

Once humans had created a sustainable food source the next step on the long road to civilization was animals according to Gascoigne (2001). The very first animals known to have been domesticated solely for a source of food are sheep in the Middle East. The proof is the high proportion of bones of one-year-old sheep discarded in a settlement at Shanidar, in what is now northern Iraq. Goats follow soon after, and these two become the standard animals of the nomadic pastoralists tribes which move all year long with their flocks, guided by the availability of fresh grass.

Cattle and pigs, associated more with settled communities, are domesticated slightly later but probably not long after 7000 BC. Humans in western Asia may first have bred the ox. The pig is probably first domesticated in China. The first reason for herding sheep and goats, or keeping cattle and pigs in the village, is to secure a regular supply of fresh meat. The hunter is dependent on the luck of the chase; if more animals are killed than can be immediately consumed, meals from the surplus will be increasingly unpleasant as the days go by. The herdsman, by contrast, has a living larder always to hand and a supply of dairy products as well.

Taking root around 12,000 years ago, agriculture triggered such a change in society and the way in which people lived that its development has been dubbed the "Neolithic Revolution." Traditional hunter-gatherer lifestyles, followed by humans since their evolution, were swept aside in favour of permanent settlements and a reliable food supply. Out of agriculture, cities and civilizations grew, and because crops and animals could now be farmed to meet demand, the global population rocketed from some five million people 10,000 years ago, to more than seven billion today.

There was no single factor, or combination of factors, that led people to take up farming in different parts of the world. In the Near East, for example, it's thought that climatic changes at the end of the last ice age brought seasonal conditions that favoured annual plants like wild cereals. Elsewhere, such as in East Asia, increased pressure on natural food resources may have forced people to find homegrown solutions. But whatever the reasons for its independent origins, farming sowed the seeds for the modern age.

Once humans had taken their first steps towards the domestication of animals and plants, land was the next valuable commodity and the one way of acquiring more resources was to go to war with a neighbouring tribe and take their resources. This meant not just conquering land but subjecting the population by means of near genocide, killing everyone that could possibly cause any future trouble for the victor's tribe or hold vengeance over them. Therefore the killing of all males and the enslaving of the women was common practice. It is from these very early and violent beginnings that humans developed a near normal state of perpetual war.

It is therefore reasonable to assume that humans started to develop inter-tribal aggression during this time of development and now 10-12,000 years later, we see the result of such early behavioural habits. Our children show signs of violent behaviour very early, typically between two and four years old, continuing throughout life, (see the later chapter on bulling for a more detailed discussion on this subject). So let's now jump from our early beginnings on the path to violence to today's environment and our more recent past.

Crime and violence

Those who can make you believe absurdities can make you commit atrocities. Voltaire (1694 – 1778)

To understand the behaviour of a criminal we have to look way back in our past to the time when demonology was part and parcel of normal life, it was thought that criminal behaviour was the result of a possessed mind and/or body and the only way to exorcise the evil was usually by some torturous means The History Learning Site (2014)

What is of interest here is that the cause of the violence and criminality was solely rooted with the individual, there was never a consideration that the behaviour was the result of social and cultural norms. Crowds would regularly gather to witness the burning, drowning, decapitation or an individual being hung drawn and quartered in front of a jeering mob.

Early Christians preached that heretics or individuals that were considered to have broken the moral, civil or church laws would receive summary sentences, which more than likely included death. For those individuals that were discovered by the church partaking in one of the seven deadly sins, the summery punishment handed out by the clergy was a condemnation to suffer a lifetime in hell. The sins and the ensuing punishment were, Gluttony: force feed rats, toads and snakes, Lust: covered in fire and brimstone, Anger: dismembered while still alive, Greed: submerged in barrels of boiling oil, Sloth: thrown into pits of snakes, Envy: thrown into freezing water, Pride: broken on the wheel. All these types of violence were sanctioned by the church. The church was considered to be the moral indicator for how followers should lead their lives.

It was during the time period when the Normans ruled the land of England that things began to change. One individual had the idea that the consequences of feuding peasants and knights was not very beneficial to the state, it would be very lucrative if justice could be nationalised and not left as it was then in the hands of the victim to exact justice. The killing of another individual, "homicide", was considered to be a loss of earnings, and when someone was killed, the victim's family would demand the payment of a sum of money from the killer know as 'blood money' or 'weregild' reputational payment demanded from the guilty person. This interfamily justice was not beneficial to the ruling monarchy, which during these times would need large funds in the chancery to fund wars and religious crusades.

Some time between our early development as warring tribes and the church led morals and sanctioned violence, human's behaviour and the reasons for violence changed to take the form of a more cultural basis for the justification of violence. Humans were still at war but now it was not so much for resources, but for religious doctrination.

This forgotten violence of our not too distant past seems so far away in our history that it has no impact on today's violence, so let's bring it a little more up to date.

Albert Bandura's Social Learning Theory (1971), put forward the view that delinquent and criminal behaviour is learned via the same psychological processes as any other behaviour: through learned and repeated exposure to rewards (reinforcements) that support the behaviour. This was in opposition to the general view at the time that implied that behaviour was the result of inner forces driving the individual such as needs, drives, and impulses, that all occurred on the subconscious level, again the individual was the source of these drivers and as a consequence that's where the popular view resided. According to his 1971 paper, Bandura believed that a man is neither driven by inner forces nor buffeted helplessly by environmental forces. Psychological functioning is best understood in terms of a continued reciprocal interaction between behaviour and the conditions that control it. He placed a special emphasis on the roles played by vicarious, symbolic and self-regulatory processes.

Observation plays a dominant role in the way these are integrated into our learned behaviour and it is Man's capacity to learn by observation which allows for large integrated units of behaviour to be learned. Observing other individual's affected reactions to both painful and pleasant stimuli, also develops emotional responses.

It could be seen as fairly clear that individuals learn through observation, however Bandura also covers the area of attention, which was discussed earlier in this book and we now know from this how important attention is. Bandura believed that one could not learn much from observation alone and that observation needed a person to attend to, or recognise the essential features of the models behaviour, this process is therefore seen as critical. Just the act of exposing an individual to a models behaviour will not ensure that the appropriate behaviour will be observed or attended to, there may well be a multitude of different behaviours being displayed and without some sort of motivation most may be disregarded. People or groups that an individual regularly associates with are likely to be the ones that learned behaviour is produced from.

Aggressive and violent behaviour will, in all likelihood, not be reproduced if the group consists of retreat monks that spend their days in peaceful meditation. If, on the other hand, a street gang is your usual hangout where drugs and violence are common place then the odds go up greatly.

Cesare Lombroso was an Italian criminologist who in 1876 put forward the theory of 'anthropological determinism', essentially the capacity for criminality was inherited, passed through birth to the next generation, you were "born a criminal". He also believed at the time that an individual could be identified by physical defects and that all criminals were male, this allowed for the label of 'savage' to be easily attached to anyone believed to be responsible for acts of criminality. He was so convinced that the features of the culprits would identify the individual "atavistic stigmata", that he comprised a list of features, this included flattened or upturned nose, low sloping foreheads, high cheek bones, large jaws, large chins, hawk like noses, large lips, shifty eyes, baldness!

Did he leave anything out? From this list the majority of the total world's population could be categorized in this way, maybe he was more accurate than anyone would concede, a scary thought! Coming up to date, let's have a look at some more recent patterns of thought on violence.

Yochelson and Samenow (2013)

A study of thinking patterns in criminals.

Aim: To understand the make up of the criminal personality.

Design: A longitudinal study using interviews that spanned over a 14 year period. The interviews were based on Freudian therapy techniques, which aimed to identify the root cause of the criminal behaviour.

Sample: 255 males from various backgrounds who had been found guilty by reasons of insanity and secured in a mental institution. Only 30 of the participants completed the interviews, and only 9 made any significant progress towards rehabilitation. Findings: Identified 52 thinking patterns that were common in the criminals.

These included:

External attribution they viewed themselves as the victim and blamed others for the situation. Lack of interest in responsible behaviour sees it as pointless. Closed thinking not receptive to criticism.

Conclusion: These 'errors' in thinking are not unique to criminals, but were suggested to be displayed more by criminals than law behaving citizens. They also put forward the theory of free will to explain criminal behaviour. This has five points to it:

1. The roots of criminality lie in the way people think and make their decisions.

2. Criminals think and act differently than other people, even from a very young age.

3. Criminals are, by nature, irresponsible, impulsive, self-centered, and driven by fear and anger.

4. Deterministic explanations of crime result from believing the criminal who is seeking sympathy.

5. Crime occurs because the criminal wills it or chooses it, and it is this choice they make that rehabilitation must deal with.

Does the criminal mind of one parent transfer via inheritance to the mind of their offspring? This has been a question that scientists and researchers have attempted to answer for quite some time now and the above does not really point us in a direction that one can be confident in!

The Construct We Call The Mind.

"Rabbit's clever," said Pooh thoughtfully.
"Yes," said Piglet, "Rabbit's clever."
"And he has Brain."
"Yes," said Piglet, "Rabbit has Brain."
There was a long silence.
"I suppose," said Pooh, "that that's why he never understands anything."

A.A. Milne, Winnie-the-Pooh

To date the brain and it's functioning process are still the subject of large amounts of research and, according to a popular myth, we only use 10% of its capacity! Wikipedia (2014) 'the 10% of brain myth is the widely perpetuated urban legend that most, or all, humans only make use of 3%, 10% or some other small percentage of their brains. It has been misattributed to people including Albert Einstein.

By association, it is suggested that a person may harness this unused potential and increase intelligence. Though factors of intelligence can increase with training, the popular notion that large parts of the brain remain unused, and could subsequently be "activated", rest more in popular folklore than scientific theory. Though mysteries regarding brain function remain e.g. memory, consciousness etc, the physiology of brain mapping suggests that most, if not all, areas of the brain have a function'.

The mind of humans is very closely related in structure and in some ways function to that of the 'Rat'. Research by Smith and Alloway (2013) at the Penn State Centre for Neural Engineering and affiliates of the Huck Institutes of the Life Sciences, detail their discovery of a parallel between the motor cortices of rats and humans that signifies a greater relevance of the rat model to studies of the human brain than scientists had previously known. "The motor cortex in primates is subdivided into multiple regions, each of which receives unique input that allow it to perform a specific motor function"

In the rat brain, the motor cortex is small and it appeared that all of it received the same type of input. We know now that sensory input to the rat motor cortex terminate in a small region of the motor cortex that is distinct from the larger region that issues the motor commands. Our work demonstrates that the rat motor cortex is parcellated into distinct sub regions that perform specific functions, and this result appears to be similar to what is seen in the primate brain."

"You have to take into account the animal's natural behaviours to best understand how its brain is structured for sensory and motor processing,". For primates like us, that means a strong reliance on visual information from the eyes, but for rats it's more about the somatosensory input from their whiskers." In fact, nearly a third of the rat's sensory motor cortex is devoted to processing whisker related information, even though the whiskers occupy only one third of one percent of the rats total body surface. In humans, nearly 40 percent of the entire cortex is devoted to processing visual information, although the eyes occupy a very tiny portion of our body's surface. It certainly seems from this research that the cortical mapping that occurs in the brain of a human is very similar to that of a rat; the big difference is the inflated size of our cerebral cortex.

Primitive neuro anatomy of the brain include impulses of rage and fear, that are balanced by the operating functions of the orbital cortex, which is responsible for emotional controls, that we know as moralization and self-control. The brain is certainly complex. However, the boundaries of its operations are slowly beginning to fail, not least due to the unfortunate circumstances some individuals have had to endure when accidental damage occurs to regions of their brain.

Pinker (2012) recounts an unfortunate accident that happened to a man called Fineus Gage, a railway foreman responsible for dynamite placement, he tapped down some blasting powder in a hole in a rock, setting off a premature explosion that sent the blasting iron up through his cheekbone and out the top of his skull. A 20th century computer reconstruction of the damage to the brain based on the holes in the skull, suggest that the rod tore up his left orbital cortex, along with the ventral medial cortex on the inside wall of the cerebrum.

Gage's sensory, memory and movement were still available to him, although something about him had changed, he was no longer the same person, the damage that had occurred had caused an effect that was not just the loss of a capability that was controlled by the brain, this was more a change in his animal like behaviour.

Pinker quotes his doctor at the time saying "he is now fitful, uses the grosses of profanities, does not care about his friends, is persistently obstinate, plans future actions which are quickly abandoned, a child in his intellectual capacity and manifestations, yet has the animal passions of a strong man. Previous to his injury he possessed a well-balanced mind and was looked upon by those who knew him as a shrewd smart businessman, very energetic and persistent in carrying out all his plans. In this regard his mind was radically changed, so much so that his friends would say, that he is no longer Gage"

This type of evidence points towards clues that the brain and the control of emotions are closely linked and interactive with each other, some parts responsible for holding other parts in check.

This leads to an understanding that the human brain has been wired for violence, it is not a random development and in our evolutionary past was required as part of human nature to ensure survival, by the use of predation, dominance and vengeance. We must also not forget that humans have a great capacity for self-control, seeking peace or loving thy neighbour. However it is these acts of violence that are really nothing other than a means to strip resources from another individual that we now term as criminality.

One particular region of the human brain that contains several different areas all linked together, and is believed to be responsible for violent acts, is a region called 'the rage circuit' The neuro scientist Yank Punck Cept describes what happens when he sent an electrical magnetic current through a part of the rage circuit of a cat! "Within the first few seconds of the electrical brain stimulation, the peaceful animal was emotionally transformed, it leapt viciously toward me with claws unsheathed, fangs barred, hissing and spitting. It could have pounced in many different directions, but its arousal was directed right at my head, fortunately a plexie glass wall separated me from the enraged beast.

Within a fraction of a minute after terminating the stimulation the cat was again relaxed and peaceful and could be petted without further retribution'. This rage circuit in the cat brain has a corresponding counterpart in the human brain cited by Pinker (2012) This region in our own brain, can also be stimulated in the same manner as the cat, eliciting emotionally enraged responses, the only difference is that the cat hisses whereas humans have a propensity to use in appropriate language and violence.

One of the distinct differences in violent behaviour is between violence that is being used for dominance and violence used for predation. Observe two cats who find themselves faced off against each other, their hair stands on end, they assume a hunched and erect posture and all manner of cat noises emanate from within, so much so that when some humans use noise as a means of posturing, we find the term 'cat fight'. Yet when the same cat comes upon a mouse or bird the behaviour is markedly different, now the cat is silent, determined and single mindedly focused on taking the life of the poor creature in its path.

Humans display the same behavioural patterns, these are evidenced in the typical Saturday night encounter when two males face off against each other. They inflate their chest, clench their fists, use language that threatens and insults the other party, however in the majority of cases even when fights start they are usually all blown out very quickly, they may have a few bruises and maybe a bone or two broken but there is, in the majority of incidents, no lasting trauma and unless they are very unfortunate to sustain a fall, and strike their head in just the right place with just the right amount of force, then death will not occur. When a tool such as a blade is involved the percentages rise sharply in favour of death.

However, we also have the capacity for predation, which unveils itself in our ugly capacity to take the life of another human in such a manner as to cause disgust and outrage. We can stalk other individuals and subject them to all manner of depraved acts eventually taking their lives. Cannibalism is also evident in some tribes and was more commonplace in our history than many would like to admit.

Humans also have the capacity to switch from passive 'I love the world and everyone in it' to 'temper enraged maniacs' at the switch of a button. This behaviour is exactly like the electrically induced rage of the poor cat above. Then we have times when humans are out for revenge, during these times a cool calculating persona can be seen, stalking their prey and preparing for the sweet taste of payback, usually a blade or a gun in some parts of the world are used in a cold manner where death is a high probability. No words are used and the silent determination is like evil unleashed.

A good friend of mine was returning home one night when he came upon a group of young lads bulling another, he intervened, trying to calm the situation, the next thing he knew and remembers was one of them repeatedly striking him, he soon went down as a result of multiple stab wounds. One thing that sticks in his mind was the coldness of his attacker executing his assault in complete silence with the rage of a person possessed.

Scientists have been able to insert their electrodes into different rage circuits within the brain of a cat to elicit either hunting or attack mode behaviour Pinker (2012). It is therefore no great leap to see that humans have the same rage circuits within their brains and that different stimuli will bring forth the same behaviour patterns that the majority of our animal relatives also exhibit.

The rage circuit that is responsible for producing emotional responses that are linked to aggression, hunting and attacking can have very subtle effects that at first look the same. These circuits are organized in a hierarchy which emanate from the 'hind brain' where neuro mapping controls the muscles and behaviour actions of the animal. The hind brain is positioned on top of the spinal cord. However, the circuits that control these rage centres are situated higher up in the mid and fore brain. When the hindbrain of a cat is stimulated by electrical impulses the resulting rage is known by neuroscientists as 'sham rage' the cat hisses, bristles and extends its fangs, but all the time can be petted and stroked without fear that the individual will be attacked.

If the rage circuit higher up is stimulated, then the resulting emotional effect is much more significant, the cat becomes as mad as hell and instantly attacks the head of the nearest person.

Evolution has, over time, taken advantage of these different modes of reactions, animals use different body parts as offensive weapons, including, jaws, fangs, and antlers, with primate's hands and feet. The hindbrain circuits that drive these peripheral actions can be reprogrammed or swapped out as a lineage evolves. The central programs that control an animals emotional state are remarkably conserved, including the lineage that leads to humans.

Neuro surgeons have discovered a counter part to the rage circuit of other animals in the brains of their patients. Pinker (2012) It would seem from these types of experiments and the discovery that human brains are not that different in their mental processes, that behavioural actions are not all under the complete control of the conscious mind and that mechanisms within our brains are pre wired for violence. Pinker goes on to describe the position and links to other systems of our brain.

The rage circuit is a pathway that connects three major structures in the lower parts of the brain. In the mid brain there is a collar of tissue called the 'periaqueductal grey', grey because it consists of grey matter, a tangle of neurons lacking the white sheaths that insulate output fibers, periaqueductal because it surrounds the aqueduct, a fluid filled canal that runs the length of the central nervous system from the spinal cord up to large cavities in the brain.

The periaqueductal grey contains circuits that control the sensory motor components of rage, they get input from parts of the brain that registers pain, balance, hunger, blood pressure, heart rate, temperature and hearing, particularly the shrieks of a fellow rat, all of which can make the animal irritated, frustrated or enraged. Their output feeds the motor programs that make the rat lunge, kick and bite, one of the oldest discoveries in the biology of violence is the link between pain or frustration and aggression.

When an animal is shocked or access to food is taken away it will attack the nearest fellow animal or bite an inanimate object if no living animal is available. The periaqueductal grey is partly under control of the hypothalamus, a cluster of nuclei that regulate the animals emotional, motivational and psychological states including hunger, thirst and lust. The hypothalamus monitors the temperature, pressure and chemistry of the blood stream and sits on top of the pituitary gland, which pumps hormones into the blood stream and amongst other things, regulates the release of adrenalin from the adrenal glands and the release of testosterone and estrogen from the gonads, which are part of the rage circuit.

In humans the Amygdala modulates the hypothalamus, as you will remember from earlier the Amygdala is responsible for memory, it also affects the emotional feeling that occur especially when fear is present and will encode these memories into our mind to remind us exactly what fear we should be tuned into. An angry face, aggressive posture, clenched fist, will all trigger neural activity in the Amygdala, this in turn sends a communication to our conscious mind with the message 'remember the last time'

At the beginning of this chapter, I laid out two categories of violence, social violence and A social violence. It is now reasonably clear that structures and mechanisms within our brain produce two basic behavioural patterns, that of predation and domination and it is these two categories that link themselves to social or A social violence. Social violence being the path to domination and the attaining of resources, A social violence the path to predation, the killing of our own species, to also enhance the attainment of resources to survive and propagate, but not always.

The reasons we construct to explain why these behaviours are enacted are our minds attempt to civilize the moral code that many now live by, whereas in an age gone by, things were very different from what they are now, the rule of law and society supported aggressive, violent behaviour in a much more open and visceral way. Yes, we have also got the capacity for great acts of kindness and compassion, we are altruistic, cooperative, but let us not be deceived by this dichotomy, for humans have evolved complex structures to ensure survival, the showing of reciprocal lateritic behaviours is just another way of banking some credit for the possibility of future hardship.

Genetics

With all this seemingly hard evidence, the question has already been partly answered and that's the question that sits just beneath the lips of people that look at their progeny and ask will they be like me, aggressive, violent, or maybe loving and caring?

Is it really the case that criminality and a propensity for violence is heritable? That's a question Raine (2013) answers in his book The Anatomy of Violence, he opens his chapter Seeds of Sin, The Genetic basis to Crime with the following recount of two lives that had never met.

Jeffery Landrigan never knew his father. He was born on March the 17th, 1962, to a mother who abandoned him at a day care centre when he was just eight months old. But little Landrigan got lucky, he was adopted by an all American family in Oklahoma. His adoptive father was a geologist named Nick Landrigan, whose wife, Dot, was a doting mother to both Jeffery and their biological daughter, Shannon. Well educated, straight-laced, and respectable, they provided a perfect new beginning for little Jeffery.

Yet an insidious shadow from the past was cast over this baby that effectively sealed his fate. By the age of two he was already throwing temper tantrums and displaying emotional dyscord that quickly escalated. He began abusing alcohol at the age of ten. His first arrest came when he was eleven, after he burglarized a home and attempted to break open the safe. He skipped school, abused drugs, stole cars and spent time in detention centres. He was moving rapidly into his criminal career. When he turned twenty he had a drinking bout with a childhood friend who wanted Jeffery to be the godfather of his soon to be child. Jeffery's response? He stabbed his friend to death outside his friend's trailer. In 1982 he started a twenty-year sentence for second-degree murder.

Incredibly, Landrigan escaped from prison, on November 11, 1989, and headed out to Phoenix Arizona. It could have been a new life and clean sheet, yet murder seemed almost destiny for Landrigan. In a Burger King in Phoenix he struck up a conversation with Chester Dyer. Dyer was later found stabbed and strangled to death with an electrical cord, with lacerations on his face and back. Pornographic playing cards were strewn around the bed, with the ace of hearts propped up maliciously on the victim's back. But Landrigan's luck was running out. While exiting the apartment he left his footprint in sugar on the floor. He was consequently arrested, found guilty of homicide and sentenced to death.

This might have been the last chapter in Landrigan's dramatic, topsy-turvy life. But the strangest twist was yet to come. While Landrigan was on death row in Arizona, another inmate told him of a man named

Darrel Hill, a con he had met while on death row in Arkansas. Darrel Hill was Jeffery's spitting image. Hill turned out to be the biological father that Jeffery Landrigan had never seen. He was a dead ringer for Landrigan and looks were not the only eerie similarity.

Darl Hill had himself started his criminal career at an early age. He too was a drug addict. Like Landrigan, he had killed not once but twice. He too had escaped from prison. Landrigan had clearly inherited much more than his father's looks. They could hardly have been more similar. And that's not all, Jeffery Landrigan's grandfather, Darrel Hill's father was also an institutionalized criminal, who was shot to death by police after he robbed a drug store in a high speed chase in 1961. He died just feet away from his twenty one year old son Darrel.

Some might argue that this is just a coincidence and there is no proof that genes have the capability to pass on a trait such as killing or criminality, yet Jeffery was raised in a completely different social environment than his father, was given all the love and support a young developing boy could ask for and still he was not able to escape the tendency for crime and murder. It would seem from this account that there lurks within our DNA makeup the capacity for inherited behaviours, other than antidotal evidence is there any research evidence that could lend weight to an inherited tendency?

Well as you would expect, Raine goes into a great deal of detail and research, one area that helps us understand the violent mind is his research into Reactive and Proactive violence. Proactive is the type of person that uses violence to extract something from someone, they are capable of planning their actions and will use varying degrees of violence and deception. They may drug their intended victim, lull them into a trap, and have sex with them all with the intention of using violence to gain a resource or to kill. These actions are carefully planned and executed and are driven by a need for a reward that can be either materialistic or psychological, they are cold blooded and have no remorse.

Reactive aggression is no more less violent than the proactive, however they do not plan ahead, instead they react emotionally to situations that get them hot headed almost instantly. Opposite to the cold blooded killers these are hot blooded, their anger bubbling just beneath the surface, only requiring the recipient of their emotion to be in the wrong place at the wrong time, say the wrong thing and set their hair trigger of violence into action.

Raines categorized 41 murders into proactive, predatory killers and reactive emotional killers. He looked into the details that surrounded each kill, looking at every piece of evidence that they could find from hearing transcripts, to interviews with the prosecution and defence team. Out of the 41 subjects, they classed 24 as reactive, 15 as proactive, with 11 remaining unclassified due to their murders containing an element of both proactive and reactive.

He then scanned the brains of each murderer and found that the reactive killers had lower prefrontal cortex functioning than a normal control subject, whereas the proactive killers had the same prefrontal cortex functioning as the control subjects. This indicates that there are other supporting biological psychological issues at play with cold-blooded killers. Raines concludes that there is a cerebral basis to violence, with the prefrontal cortex playing the major part in the separation between cold or hot-blooded killers.

Pinker gave us the "rage circuit" and puts forward evidence that we are all in some way wired for violence, if this is the case then Raine's research underpins the fact that some humans have brain markers that make them susceptible to either reactive or proactive violence, supporting a general genetic ability for some type of violence.

Raines, conducted extensive research into individuals who had a low resting heart rate and their propensity for violence. His findings indicate that a low heart rate is a significant indicator for these individuals to be at risk of behaviour that is A Social and violent. He also goes on to link other forms of maladaptive behaviour of the mother and social climate, in which children are raised. Mother's smoking, social deprivation, poverty and poor nutrition, all are contributing factors that together with a genetic and evolutionary marker all go to support the clear fact that the race homeo sapian is a violent one.

The overall goal of this chapter was to take the reader through an evolutionary time journey into the reasons behind why humans are violent. How we, as a race, achieved such fantastic things in the face of such an overriding aggressive trait? Make no mistake, we are wired with a genetic code that allows us to carry out the behaviour of violence, but we also have the capacity for great feats of kindness and love. To operate within the world of violence and still maintain a connection to all that is good within humans, takes a clear understanding of why we are the way we are.

Knowing these facts can help develop behaviour that enables easy access for violent encounters at the switch of a button, but also, and in my view more important, allow us to be at peace with ourselves, both during and after any act of violence. As long as your moral ethics are solid and support your actions for violence, then the cost in psychological and emotional states that are harmful to your wellbeing can be alleviated.

7 NEURAL MUSCULAR PROGRAMMING

"Everything should be made as simple as possible, but not simpler"

Albert Einstein

In today's combative/martial arts, there seems to be a change in focus towards the arts that profess to be reality based, they use descriptions such as street effective, functional, the real art, to name but a few, there are even arts out there that have the title scientific!

What I see is a shift towards intelligent arts using up to date thought processes to base teaching on, rather than the old way, which was do as I say. If any questions were raised the answer was all too often hidden within the mysticism of the arts. The new systems are changing, they are asking questions and demanding logical well thought out answers that have a sound thought process. The universal problem is how do you know? How do you know that a particular training method is adaptive rather than maladaptive? As we have discovered throughout our journey through this book sometimes even the best get it wrong!

Here's a question, what level do you want to get to within your training? I do not mean what level of grade, I mean what level of execution and performance, and what level of effectiveness do you want? In chapter 9 I explore performance under stress in greater detail, what you will find is a link between Neural Muscular Programming (NMP) and performance.

There are three basic levels of application of movement that apply to most physical executions of a skill-based activity. When any individual starts learning a new skill based activity it's going to take a while before they are able to perform their moves with any level of competence, to begin with, there is no real understanding of why they are doing certain moves, they are just getting to grips with a new way of moving. This can sometimes take years, during this time the type of movement that they perform could be categorised as Mechanical Movement (MM). They have learnt the moves and can perform them as required by their instructor. It is during this early stage when it is very important to ensure that all moves are questioned and critically anaylised. If the art that you practice does not encourage questions and logical answers, you could be stuck doing mechanical movement for a very long time. Never learning why moves are effective and why some are not, never understanding how to produce power or speed at its highest level. I am not saying that the mechanics of movement are not effective, it is a matter of where on the scale do they come.

There is, non-effective, effective, more effective and most effective, and yes, everyone thinks that what they do is the most effective method, nobody is going to be able to withstand the techniques that you have learnt, then one day the penny drops, your mind starts to think in a different way, the filters that you build instinctively begin to let in a new thought process. "To work properly the mind has to be like a parachute, unless it's open it's not working efficiently"! Parker (1990) Our conscious mind is always working at below its full capacity, it filters information and only feeds you what it thinks you can deal with, if it didn't, we would all soon freeze, just like are computers!

To be able to move through different levels within the combative/martial arts we need to be able to think and to understand what we are doing and why we are doing it, this then helps us to move to the next level which I will label Engineer of Movement, [EM]. This level is where we find the majority of instructors and long term students of the arts. They have the answers to your questions and can coach you towards a deeper understanding and a much higher effectiveness in how you are moving. The mind is able to call upon the moves and techniques that it knows are going to work given your current situation.

Not only can the EM move well, he can explain the reasons behind why he is doing certain things. He is also able to go back to the drawing board and come up with a way around any problems that may occur. This is due to the fact that he understands the basics and all the specifics. He knows how and why things are done and can therefore re-work around the problem. This is where your instructor should be and it should be your first target as a practitioner. The mind has learnt the movement, it understands the principles that make the movement effective and it allows access at any time, it has built new filters and programmed itself to a much higher level of execution than can be obtained by the MM.

So what's the highest level?

My thoughts on this are that very few people can ever work at the highest level, most of us access this level some of the time, hopefully when we need it most, when we are in real trouble!

This level is the 'Magician of Movement' (MoM). This is when a person moves effortlessly with no exertion, their technique is the most effective with the minimum effort, they are able to constantly adapt and change as the situation requires, there is no pre-thought, no plan, techniques merge into one another, the slightest touch is as powerful as it can be, they move with such speed that it's impossible to track their movement. The mind has processed everything that they have been taught and learnt and is now processing this on a sub conscious level. This is the realm of the 'Magician of Movement'!

This is where true spontaneity can be found, after all what is your goal? Do you want to have to consciously think before you move? I think not, that's going to be slow, you need to be able to react without pre thought, you want to get as close to that reflex time of 19 - 24 milli seconds (ms) as we can and certainly quicker than 300 ms of Conscious Reaction Time (CRT).

The difference here is the way the mind is accessing and processing information and to get here you need to go through all the previous levels, you need to start thinking in different ways and this is where NMP comes into the picture. Once you have been through these levels the next stage is to abandon your conscious thought and let your mind do the work, once it has been programmed to such a high degree it is absolutely natural for it to do the work without pre thought, after all, your autonomic nervous system is always working at this sub conscious level.

NMP within the combative/martial arts is a new term, people are beginning to use the terms NLP and once they understand NMP it will also be used. Let's explore the link between mind and body and how it relates to the combative/martial arts.

Somatosensory System

To understand NMP we first have to look at the senses within the human body and explore one particular sense in a little more detail. For many years the majority of school lessons revolve around teaching children that we have 5 senses, touch, smell, hearing, vision and taste. Today's science has come up with a few more and depending upon your viewpoint this could be as many as twenty-one. The sense that I am interested in here is controlled by our somatosensory system. This system perceives, relays and processes somatosensory sensations, like touch, vibration, pressure, itch, nociception (information about painful stimuli), temperature, proprioception (information about the position and movement of our joints and muscles) and interoception (information about internal organs). The one that is of real interest is that of 'proprioception' although nociception is also important when is comes to understanding body reactions to pain and touch, vibration and pressure all link into NMP.

Modality Specificity in the somatosensory system can be sub divided into several different categories of perceived sensations i.e., pain, temperature and touch proprioception, which can also be divided again, but let's not get too technical here. One category that has value to NMP is discriminative touch, which is divided into pressure, flutter and vibration. To provide a quick example of proprioception and the NMP link I will discuss how the formation of the hand posture, the linking and touching of the fingers relay certain information back to the brain. The idea of postural references that created a change in the application and efficacy of movement was first introduced to me by Chapel in 2005 during lessons and seminars, although at the time not referred to as NMP, its application to human movement in this way certainly supports this theory.

As an example of an attack, let's take a headlock technique, an attacker attempts to attack you in this manner with their right arm around your head, you managed to grab their arm.

What posture have you created and what are the proprioceptive senses relaying to your brain? Your intention could be one of two actions at this point, 1, to pull their arm away from your neck relieving pressure, and 2, manipulate their arm. The posture that you create will automatically relay proprioceptive information back to your brain. You have grasped their arm, your thumb is wrapped around one side of their wrist and the fingers are on the other side. Are you working at your most efficient if your intention at this point is to pull down to relieve the pressure? This posture has created a dichotic relay of information to the brain from what your intention initially was. This particular posture is communicating an entirely different application through the proprioceptive senses. When the posture of the hand is as described above, instead of a pulling action the senses that have been programmed in are communicating a grasping action, one where your intention is to manipulate their arm, controlling it.

If you were to hook all your fingers around the arm and leave your thumb out of the posture, you will now have formed a posture with your hand that is communicating to your brain that you are trying to pull something. The difference here is very specific, forming the correct postures that match your mental intent is a key factor to working at your most efficient, you have linked what your body postures are telling your brain into what the mind instinctively thinks your body is trying to achieve. You are now performing your moves at your optimum level. You have created the NMP link, all the filters and links between body and brain are working as one, and these filters are aligned. Now you have relieved the pressure from your neck with the pull posture, you are now free to change the posture into the grasping one, which then enables you to manipulate the arm.

I hope that this very specific example helps to explain why it is so important to have the mind and the body programmed to work in the correct way and not out of sync, as so often happens in the combative/martial arts, when we have trained our body to work inefficiently!

The somatosensory system has been associated with a sense of self-awareness, as this system is providing multiple feedback loops, which allow the brain to know where the body is in space and time and to relay a sense of internal representations of the body. Just like many of the other systems that are contained and managed within the cerebellum the somatosensory system has to make literally millions of calculations from moment to moment just to maintain the balance of the human body.

Neuro Muscular Programming, as a term was first used back in the Eighties by cognitive neuroscientists, who believed that the brain could be accessed and altered, as with the neuro-plasticity already discussed, and that with the somatosensory proprioceptive feedback senses and the power of mental attention, the brain's control unit could be accessed and changed. Also the use of already programmed synaptic pathways between body posture and movement could be used to enhance efficiency. Proprioceptive sensations trigger reflexes that are also used to protect the body from injury as well as providing feedback on the position of the body as it relates to itself.

Over our lifespan we use NMP every day to perform their natural movement. A connection is built over time and is continually our reinforced every time we perform an action, our brain recognises the postures and muscle-firing sequences that tell it what action is intended by the body. The real trick is to create links from NMP to combative/Martial Art movement.

Proprioception

Proprioception is the sub conscious perception of movement, position, location and spatial orientation of the body and its parts arising from stimuli within the body itself. There are various specific mechanoreceptors which respond to the stretching of a muscle, a tendon or a ligament or respond to the movement of special structures in the inner ear.

Dougherty (2014) of the Department of Anesthesiology, Pain and Medicine, at MD Anderson Cancer Centre provides information regarding proprioceptors on the website Neuroscience Online. Proprioceptors are located in muscles, tendons and joint ligaments and in joint capsules. There are no specialized sensory receptor cells for body proprioception. In skeletal (striated) muscle, there are two types of encapsulated proprioceptors, muscle spindles and Golgi tendon organs, as well as numerous free nerve endings. Within the joints, there are encapsulated endings similar to those in skin, as well as numerous free nerve endings. It's this system of sensory feedback located throughout the whole body that communicate position and postures back to the cerebellum, which in turn enable the mind to decode the information and know what the body is up to, hence the hand posture detailed above seems to now make a lot more sense. The link between the body postures and specific movement sensed by the proprioceptive system and relayed to the cerebellum is ultimately the mechanism for creating NMP. Continued repetition over our lifetime programs the motor cortex, registering what actions or movement is intended by the body's actions.

Evidence of this programming of the motor cortex has been found when the body fails, usually due to injury. What has been discovered is that even in older individuals the brain has the ability to reorganize cortical representations. To provide some background on this I want to look at some research that has been conducted on stroke patients. For centuries it was a given fact that there was no plasticity of the adult brain and the axiom based on Hebian remodeling "neurons that fire together wire together" was all there was, cortical plasticity was something that just did not happen.

Back in 1931, Sherrington conducted research on sensory deafferentation and concluded that this procedure led to a loss of motor ability, which led to the modulation of reflex pathways, which is the basis for compulsive behaviour. The brain's motor cortex taps into pre-existing motor cortex circuitry to carry out its commands, in other words all volitional behaviour is built on integrated hierarchical reflexes. An animal moves and this movement produces sensory feedback, feedback plus learning guide the next movement, this produces its own feedback which again, in conjunction with learning, eventually produces after countless iterations, purposive sequential movement. This theory came to be called Sherintonian reflexology cited by Schwartz (2002). This cycle of sensory feedback, which involved volitional attention to effect the motor pathways through repeated repetition was the first work that began to unfold the connection to NMP although at the time of this research, neural plasticity was a subject that attracted very little attention due to the un-ubiquitous support of neuroscientists of this time.

What we know today about the brain's capacity for cortical reorganization was an unexpected by-product of research conducted by Edward Taub (1931) who was a behavioural neuroscientist, he is best known for his research on the Silver Spring monkeys back in (1981). Although this research at first did not have the outcome that anyone would have imagined Wikipedia (2014). The Silver Spring monkeys were 17 wild born Macaque monkeys from the Philippines who lived inside the Institute of Behavioral Research in Silver Spring, Maryland. From 1981 until 1991, they became, what one writer called, the most famous lab animals in history as a result of a battle between animal researchers, animal advocates, politicians, and the courts over whether to use them in research or release them to a sanctuary. Within the scientific community, the monkeys became known for their use in experiments into neural plasticity, the ability of the adult primate brain to reorganize itself regarded as one of the most exciting discoveries of the 20th century.

The monkeys had been used as research subjects by Edward Taub, a psychologist, who had cut afferent ganglia that supplied sensation to the brain from their arms, then used arm slings to restrain either the good or deafferented arm to train them to use the limbs they could not feel. In May 1981, Alex Pacheco of the animal-rights group People for the Ethical Treatment of Animals (PETA) began working undercover in the lab, and alerted police to what PETA viewed as unacceptable living conditions for the monkeys.

In what was the first police raid in the U.S. against an animal researcher, police entered the Institute and removed the monkeys, charging Taub with 17 counts of animal cruelty and failing to provide adequate veterinary care. He was convicted on six counts; five were overturned during a second trial, and the final conviction was overturned on appeal in 1983, when the court ruled that Maryland's animal cruelty legislation did not apply to federally funded laboratories.

The ensuing battle over the monkeys' custody saw celebrities and politicians campaign for the monkeys' release, an amendment in 1985 to the Animal Welfare Act, the transformation of PETA from a group of friends into a National Movement, the creation of the first North American Animal Liberation Front cell, and the first animal research case to reach the United States Supreme Court. In July 1991, PETA's application to the Supreme Court for custody was rejected, and days later the last of the monkeys were killed.

During the subsequent dissection of the monkeys, it was discovered that significant cortical remapping had occurred, suggesting that being forced to use limbs with no sensory input had triggered changes in their brains' organization.

This evidence of the brain's plasticity helped overturn the widely held view that the adult brain cannot reorganize itself in response to its environment. After five years of receiving death threats and being unable to find a research position, Taub was offered a grant by the University of Alabama, where he developed a new form of therapy based on the concept of neural plasticity for people disabled as a result of brain damage.

Known as Constraint-Induced Movement Therapy, it has helped stroke survivors regain the use of limbs paralysed for many years, and has been hailed by the American Stroke Association as at the forefront of a revolution. In chapter one I explored the power of attention in creating a recognizable change in the ability to adapt and change, this research supports this view. This leads to an understanding that the power of thought, the ability to pay attention has a causal effect on behaviour and our mental states, which in turn causes cortical re-organisation and plasticity to occur.

Although at first the above research seems to have been completely intended for other use, it did not take long before the benefits to humans became apparent and the fact that although the brain is programmed in a certain way it is not immutable.

This leads us on to a large body of research based upon human cortical reorganization as a result of an injury to the brain. To be locked into the idea that NMP is a hard wired concept would, given the research, be a mistake. What this does support, is that teaching skills and mental thought processes in the correct manner could have huge consequences for the way movement is trained and programmed into our brain.

The following research below takes a look at how plasticity affects individuals that have had cortical reorganization due to injury and was conducted by Liepert, J. MD. Bauder, H. Ph.D. Miltner, H. Ph.D. Taub, E. Ph.D. and Weiller, C. MD. (2000). They state "there is almost no information on treatment-induced plastic changes in the human brain". Their aim was to study and evaluate reorganization in the motor cortex of stroke patients that was induced with an efficacious rehabilitation treatment. They used focal transcranial magnetic stimulation to map the cortical motor output area of a hand muscle on both sides in 13 stroke patients in the chronic stage of their illness before and after a 12 day period of Constraint-Induced Movement Therapy.

What their research revealed was that before treatment, the cortical representation area of the affected hand muscle was significantly smaller than the contra lateral side. After treatment, the muscle output area size in the affected hemisphere was significantly enlarged, corresponding to a greatly improved motor performance of the paretic limb. What they discovered was that plasticity occurred once attention and repetition of movements were undertaken.

The reorganization of further regions within the brain indicates that a recruitment of adjacent cells was taking place, the brain was being re-mapped. When they conducted follow up examinations after 6 months of rehabilitation exercises, motor performance remained at a high level, whereas the cortical area sizes in the 2 hemispheres became almost identical, this indicates that not only did the brain return to a balanced cortical use within each hemisphere, but also that as both sides of the body were being used a bilateral symmetry was occurring in the control systems.

Their research evidence was the first demonstration in humans of a long term alteration in brain function associated with a therapy induced improvement in the rehabilitation of movement after neurological injury.

The research by Sherrington and later Taub on animals led to the discovery that cortical reorganization occurs after injury to the nervous system and that with the right circumstances recovery of motor actions were achieved. This consequently led to a theory that spontaneously occurring cortical reorganization phenomena that result from nervous system damage or conditions that involve abnormal sensory input have been shown to be associated with pathological states in humans; these include phantom limb pain, tinnitus, and focal hand dystonia. After motor stroke, a complex pattern of reorganisation has been described.

Within this book it has been my intention to provide evidence that certain types of training can be conducted that are very much more in-line with the natural processes of the human body's ability to adapt and alter its movement and that maladaptive behaviour can easily be programmed into the brain of an individual.

This programmed information on movement is critical, as getting it wrong could have disastrous consequences. NMP is achieved by the continuous repetition of movement; this is supported by the cortical reorganization that has been found to occur within the brain. However, evidence was also seen in the preceding chapters that indicated that use dependent cortical reorganisation also occurs and that attention driven volitional force has a causal effect on the brain, and continual use of a body part or parts in behavioral relevant tasks results in an increase in cortical representation onto the cerebral cortex.

Not only can this process help in the development of combative/martial movement, it could also be used in the rehabilitation of individuals that have damage to critical areas of the brain, like those stroke patients studied above. Indeed, therapy has already began to be developed and targeted at stoke patients.

Constraint Induced Movement Therapy

Liepert, J. MD. Bauder, H. Ph.D. Miltner, H. Ph.D. Taub, E. Ph.D. and Weiller, C. MD. (2000) Go on to discuss Constraint-Induced Movement Therapy (CI therapy). They show that this therapy has been observed in controlled studies to be efficacious in chronic stroke patients.

The following is a brief look at their research.

Patients with chronic strokes that have been at this stage for a period of time are presumed to have a stable motor deficit. Moreover, the short duration of CI therapy (12 days) further minimizes the possibility that spontaneous recovery of function could give the appearance of a treatment effect. CI therapy stems jointly from basic research in neuroscience with monkeys with somatosensory deafferentation of a single forelimb and from behavioural psychology. The effective therapeutic factor in this treatment appears to be the massing or concentration of practice in use of the extremity affected by a stroke for many hours a day during a period of consecutive weeks.

This therapy has been found to produce a substantial long term improvement in the amount of use of an affected upper extremity that transfers into the real world environment. It is possible that CI therapy might produce its therapeutic effect through the induction of a use dependent cortical reorganization that counteracts adverse brain function changes and enhances recovery associated plastic changes that occur in the human brain after stroke.

The main goal of the present study was not to evaluate the clinical effects of CI therapy or to compare this treatment with other physiotherapeutic approaches but rather to use CI therapy as a model to assess therapy induced plasticity in stroke patients. Therefore, they did not introduce a control group. However, they did use a control procedure (i.e., 2 complete pretreatment test batteries separated by the same length of time required by the intervention), placebo controls have been used in other CI therapy research.

Focal Transcranial Magnetic Stimulation (TMS) was used to assess plastic alterations that may have been induced by CI therapy. TMS involves the non-invasive mapping of motor regions of the brain to determine the cortical representation areas of muscles with the use of a focused magnetic field to stimulate loci in motor areas from points on the scalp. It has been used to assess the amount of reorganization of motor representations consequent to injury of the peripheral and central nervous systems and after various conditions of use. The amplitude weighted centre of the TMS map of a hand muscle corresponds closely to the hand area within the primary motor cortex as determined with anatomic and functional MRI (fMRI) studies.

In contrast to typical fMRI or Positron Emission Tomography (PET) experiments in stroke patients, TMS mapping is performance independent and therefore ideally suited for longitudinal studies as in rehabilitation of stroke, where motor ability may change markedly. Preliminary results from a limited sample of patients had indicated that motor cortex reorganization occurs immediately after CI therapy. In contrast to this earlier study, they performed TMS mappings and evaluations of motor functions in parallel at several time points before and after CI therapy to investigate the stability of the baseline and to determine short and long-term effects of the therapy on the functional organization of the primary motor area of the brain in relation to clinical recovery.

Subjects and Methods Thirteen patients (10 men; mean age 56.7±10.3 years, age range 33 to 73 years; duration of hemi paresis 4.9±4.7 years [mean±SD], range 0.5 to 17 years) with chronic stroke (>6 months) were studied. Eleven of the subjects had a right sided paresis; 3 had cortical lesions (2 ischemic and 1 hemorrhagic in origin), and 10 had lacunar sub cortical lesions that involved the internal capsule. Informed consent for participation in the study was obtained from all patients. The study was approved by the local ethics committee.

Functional inclusion criteria were (1) the ability to extend 20° at the wrist and 10° at the fingers and (2) sufficient stability to walk when the less affected arm is immobilized. Exclusion criteria were (1) serious uncontrolled medical conditions, (2) global aphasia or cognitive impairments that might interfere with understanding instructions for motor testing, (3) anything in the head that contained metal, (4) pregnancy, (5) epilepsy, and (6) cardiac pacemaker.

Each subject received 12 days of CI therapy preceded and followed by periods in which electrophysiological, neurological, and behavioural testing was conducted. For CI therapy, subjects agreed to wear a resting hand splint secured in a sling that prevented use of the non-paretic upper extremity for a target of 90% of waking hours. This arrangement induced greatly increased use of the paretic arm. In addition, on the 8 weekdays during the treatment period, the subjects came into the laboratory and received 6 hours per day of training in use of the affected arm in a variety of tasks according to a behavioural technique termed "shaping."

The shaping was designed to produce intensive use of the more affected extremity, while at the same time, improving the quality of movement. Treatment efficacy was evaluated with the motor activity log (MAL), which tracked arm use in 20 common and important activities of daily living (ADL) performed outside the laboratory (1) for the week before the subject's visit to the laboratory 2 weeks before the beginning of treatment, (2) for the week before the beginning of treatment, (3) 1 day after treatment, and (4) 4 weeks and (5) 6 months after the end of treatment (follow-up).

The MAL has exhibited excellent interest reliability for chronic stroke patients across a 2-week interval equal in length to the treatment period when no treatment was provided and when a placebo treatment was administered. Further details of the intervention and testing are available elsewhere. They mapped the cortical output area of the abductor pollicis brevis (APB) muscle of the more affected and less affected hands with TMS on the same day as the MAL was administered 1 day before treatment and 1 day and 4 weeks after treatment.

(Two subjects died before the 4-week post treatment testing could be carried out, and 1 subject had to be excluded at this time because of the intervening occurrence of an epileptic seizure.) As a control procedure, to confirm the stability of the electrophysiological and behavioural measures, 10 subjects were tested 2 weeks before the beginning of treatment.

This is the same temporal interval that separates the second pretreatment and post treatment tests and therefore controls for such nonspecific effects as expectancy of improved extremity function, initial contact with experimenters, and increased attention to use of the affected upper extremity. During the period between the testing 2 weeks and 1 day before treatment, there was no contact between patients and the project. Eight subjects have been tested 6 months after treatment to date.

The most salient result of the present study is the almost doubling of the excitable cortex, yielding responses of a muscle in the more affected hand of patients with chronic stroke after CI therapy. This result is paralleled by the large improvement produced by this intervention in the same subjects in the amount of use of the more affected extremity in the real world setting. The behavioural and electrophysiological changes were consistent across individuals, with both being observed in each patient.

Their present study demonstrates that CI therapy has a parallel effect in humans after stroke. Similarly, in recovered stroke patients, a large lateral extension of the brain area that is active during finger movements was found. Their results suggest that a reorganization occurred on a cortical level. However, the results do not permit exclusion of the possibility that additional plastic changes occurred on a sub cortical or spinal level.

CI therapy is predicated on the demonstration in deafferented monkeys after neurological injury that the non use of an affected extremity can be due to a learning phenomenon that involves a conditioned suppression of movement. CI therapy is considered to be effective because it increases the motivation to use the extremity and thereby overcomes the "learned non use." (This formulation has been described in detail elsewhere.

The current results indicate that the intervention, which involves massed and sustained practice of functional arm movements, also produces a massive use-dependent cortical reorganization that may provide the basis for the long-term persistence of the treatment effect for the 6 months studied in this experiment and for the 2 years reported in other research.

Other examples of use-dependent cortical plasticity, resulting from the increased use of body parts in behaviour relevant tasks, have been described in animals and humans.

One of the aims of neuroscience has been to generate effective new rehabilitation strategies that would give pragmatic importance to this area of basic research. Moreover, if a central nervous system correlate of such a therapy could be found, a new area would be opened in which further improvements in rehabilitation might be produced through manipulation of that correlate. The present study addresses both of these objective.

Finally, before I leave this subject, it is important that we look at one further area of research that Taub did. Remember, according to Sherrington, movement was dependent upon modulation of reflex pathways and is the basis for compulsive behaviour, volitional behaviour is built on integrated hierarchical reflexes, an animal moves and this movement produces sensory feedback. Feedback plus learning guide the next movement, this produces its own feedback which again, in conjunction with learning, eventually produces, after countless iterations, purposive sequential movement.

Taub thought that this was not the case and continued to test his theory that the basis for movement was indeed hardwired into our genetic blueprint. He did this by conducting deafferentation research on infant and fetal monkeys and what he found was that the monkeys were able to use their limbs to feed and move themselves around, this led him to conclude that volitional movement was not dependent upon sensory feedback like Sherrington had first thought, and that is was pre-loaded into an animal's brain like 'Windows XP' on a computer, what this developed into was a theory of learned non-use Schwartz (2002).

Learned Non-Use, How Habits Form

Habits are a psychological pattern called a "habit loop," which is a three step process. First, there's a trigger, a certain stimuli that sends a message to your brain, the resulting action "the habit" then unfolds in an automatic behaviour pattern, a little like a fixed action pattern, once it's triggered it has to continue until completion.

The second step is the behaviour itself, this could be anything from a fidget to a complicated set of movements strung out over a period of time, driving a certain way to work each day for example.

The third step, is the reward for the behaviour, this again does not need to be something tangible just something that supports the "habit loop" in the future.

According to Charles Duhigg (2013), an investigative reporter for The New York Times, who has written a book, "The Power of Habit: Why We Do What We Do in Life and Business." Cravings are what fuel a habit and enable the habit loop to work.

Neuroscientists have mapped our habit making behaviour to a part of the brain called the basal ganglia, which is also responsible for the emotions, memories and pattern recognition. While our decisions emerge from the prefrontal cortex. We know from the previous chapter on attention that the brain uses less energy while working on autopilot and that's exactly what happens when a habit forms, we work on a powered down amount of energy.

Duhigg also goes on to state "You can do these complex behaviours without being mentally aware of it at all," he says. "And that's because of the capacity of our basal ganglia: to take a behaviour and turn it into an automatic routine."

What does this all have to do with effective, efficient movement or NMP within the combative/Martial arena? It provides sound evidence that the human brain is plastic and can be programmed with repetitive movements, and unlike an injury, when most people take up some sort of training within the arts, they are doing so out of their own volition, or they are in an occupation that requires physical training to carry out their duties. They are purposefully and willfully following the instruction of their teacher, beginning the process of use-dependent cortical reorganization, this at a time when their bodies have, over their lifetime, created NMP links to behaviour that has been hardwired into the blueprint of their movement.

Make sure you know what you are getting into.

How many times have you heard this? "That's wrong! this is the correct way". It can be packaged up whatever way you like, in the end it comes down to an instructor stating that they know the way that it should be done and you are the student, so your role is to follow and mimic the moves exactly! You are certainly not encouraged to question the instructor. If you are a professional entering an occupational based training, I understand the reasons why open questions are not encouraged, however when you are choosing a place to study, the environment may be a little different, so the question "why do you do the move that way"? should be asked?

All too often the reply is, "because that's the way my instructor taught me". On the face of it this is not a problem, you may think to yourself maybe he will give me the answers later. You put complete trust in your instructor and the Art that you have chosen to partake in, this in general is not a problem as long as you are aware of the reasons for your choice and the goals and direction that your chosen art follows. You may have chosen a traditional art which requires hours and hours of focus on one move, with an absolute dedication to the historical application of your moves. If this is the case, then all those hours of repeated repetition are going to create a use-dependent reorganization of areas within your cerebellum that are responsible for motor movement. Or maybe your intention is to choose one that has a focus on spiritual understanding, traditional Chinese or Japanese weapons, street self-protection, the list goes on. What's important is that your instructors are clear with regard to the arts application. The one thing you do not want is an art designed for the ancient battlefield trying to cope with the yob down the local hostelry. All I am saying is be clear on the reasons for starting the art you choose, keep an open mind and always ask questions, keeping in mind the evidence that has been included in this chapter when it comes to some of the movements that may be taught. If we understand the reasons behind the teaching, it makes it easier to accept being told "that's wrong, this is the correct way" when practicing a more traditional art.

The essence of the art itself is the discipline and the self-control needed to emerge yourself completely in mastering the exact movement required. This type of practice is arguably the hardest of all, especially in today's environment of quick fix sensory input where students quickly become bored and want to move on to the next part of the technique.

Being continually corrected on the smallest of detail soon tests the patience of the student. However continued correction and alteration is a pre-requisite of accurate mapping onto your motor cortex. It is very important to understand that if motion is repetitively practiced in a fast manner, then there is going to be quite some lag until movement becomes spontaneous, as fast repetitive movements are the hardest to ensure the neural-pathways are mapped accurately each time. The slower the practice the more likely the action will be repeated accurately and therefore become encoded into your motor cortex, again we come back to the saying "repeated accurate practice slowly makes perfect".

Being manipulated and controlled within a set perimeter of techniques, without the room to alter or adapt might not be the best way to achieve this. Remember the statement by Bruce Lee "learn what is useful and discard the rest" how do we know what is useful and what should be discarded? Especially if we are constantly told that the way we are moving is wrong and this is the right way. In a great deal of schools you can hear the words "we will teach you what comes naturally". So the question to your instructor should be; what is natural movement and how do we know it's natural?

Observe a new born child a few days old, when a loving parent places their finger onto the baby's palm, you see the baby grasp the finger. As soon as the new born is able to stand, we see the beginnings of their attempt to walk! None of this is taught, it's simply natural movement. According to Taub it's hard wired in as a process that everyone will travel through, this is the starting point for what I call Neuromuscular Programming (NMP). Over the remainder of our lives we will use NMP every day to perform our natural movement. A connection is built over time and is continually reinforced every time we perform the action, our mind recognises the postures and muscle firing sequences that tell it what action is intended by the body. It's fairly obvious by now that any training should follow the rules of behaviour and movement that we learn instinctively from the day we are born so before we move any further let's take a look at what the body can learn that is not natural, this is learned bad behaviour, which is very similar to the learned non-use that Taub discovered with the Silver Spring monkeys.

Learned Bad Behaviour

In the chapter on Bilateral Symmetry I explored the capability of humans to learn and move in a wide variety of ways, some of which are unnatural. I used walking in a straight line as an example and asked you to analyse your own walking gait. The effects of bad learned behaviour or learned non use allow the feet to fall outwards, this then transfers to the feet and subsequently your own walking gait. This area could also be termed habit forming, which is the same as the habit loop discussed above.

Which moves within the combative/martial arts do you think are natural and which ones are learned bad behaviour?

What about, a hand-sword, a punch, a step, a kick or a block? Is it natural for us to fight at all?

In the chapter on Symmetry I used the handsword as an example to look at natural weapons and the formation of postures. Remember the weapon is usually formed by touching the fingers together and straightening the fingers of the hand, so that it looks like a sword, hence the name handsword. I then provided two example of how the posture of the hand could be formed and the potential effects of the different postures were explored. What has happened during the formation of this posture is that NMP has kicked in. When the fingers touch, the somatosensory category of discriminative touch, which includes pressure perception has relayed information to the brain, telling it that the fingers are touching together, this translates into pressure and engagement of muscles within the hand to maintain this posture.

As a martial practitioner we then go and hit something hard with the side of the hand and discover that it's not quite as effective as when we open the fingers and spread the metacarpals within the hand. It is this understanding of the communications that are occurring throughout the body that enable a study of the weapon and the mental processes that back that weapon up to be built on a scientific understanding, rather than have some vague relationship between a description that somehow relayed the way that the weapon was formed and the intended use of that weapon.

All too often individuals move in a way that has been learned or is a consequence of non-use in the correct manner and it's this non-used correct method of body mechanics, that over time, affects the mechanics and overall posture of the body. We see evidence of this in surgery to correct bad knees, hips, by observing bad hunched postures, the way individuals walk and ultimately the way some individuals have been taught to move by combative/martial art instructors.

So we now know that humans have a plastic brain that can re-map the neural networks to support limbs or areas of the body that were controlled by parts of the brain which have been subjected to injury. We also know that repetition of movement will, over time, create a hard wired link between the brain and the body, resulting in cortical re-organisation. NMP is programmed in from birth and how we move or form our postures has either a negative or positive effect on our intended actions. Armed with this knowledge, we are in a position to understand movement and more importantly, ask questions of ourselves and those we entrust to teach us, with a volitional mind we have the capacity to pay attention, train our movement, altering the most complicated communication device known to man, the human brain.

8 STARTLE REFLEX I DIDN'T JUMP

Before I explore the area of startle reflex, I think that it's appropriate that I set the scene and discuss human movement as it relates to response times that are linked to areas of research around commonly used criteria for the teaching of some combat and martial arts systems, where research and laws are used to exclude other forms of movement based teachings, specifically Hick's law ("HL") Law and Power Law of Practice ("PLP"). This research used to support underlying technique, and it's important to ensure that we have all the relevant facts and not just the ones that suit the purpose of the individual, as it is so often the case that only proportions are used that support the theory. Having a context to base this information on will help understanding, it's also important to realise that in some instances we are talking milliseconds rather than seconds.

One of the goals that I want to achieve with this book is to provide solid evidence to the methods of teaching that should be employed, without having to justify teachings on information that is either outdated or irrelevant to today's theatre of combat and, although some topics will be repeated, I do not make any excuse for this as I feel that it's important to repeat some things! Many of the reality based self defence instructors use a method that has been pitched at a very specific category of violence, namely the A Social level and not the every day social violence that we encounter 98% of the time. I use this high percentage to get across my point that this is not for your every day classes that teach self defence, and therein lies one huge psychological problem, because the majority of RBSD methods believe that what they teach will allow an individual to manage and cope with A Social violence, using different situations in different environments to convince individuals that what they teach is the real thing! The "real thing"?, according to whom?

Hick's Law

Ockham was a 14th century English philosopher who first proposed the principle that "plurality should not be posited without necessity" and It's from this very unobtrusive start that we later arrive at Hick's Law and then subsequently we find RBSD instructors advocating that human movement, within a combat situation, should be trained only to a very limited amount of moves. According to Jefferys and Berger (1999) it's unclear as to what was meant by this statement, as it can be interpreted in many ways. However, later versions were clear and here is an example given by Jefferys and Berger: "entities should not be multiplied without necessity" or "it is vain to do with more what can be done with less" and finally, a more up to date rendering, "an explanation of the facts should be no more complicated than necessary". I think I may have fallen foul of this rule a few times! However it's important that information is validated and not just thrown out there to see who picks up on it, it's due to this validation that lengthy text are required along with research to provide a more clearer view.

Over the years, many noted individuals have used this theory to reduce complicated ideas to a simple more logical theory and this is all well and good when it relates to simple ideas and is used as a rule of thumb. However, humans have made great leaps forward since the 14th century, in our understanding of DNA for example, not a subject where corners could be cut to aid understanding. It's therefore easy to see how those with a limited arsenal would want to use such terms to build a self defence system upon. This theory was then backed up years later by Hick and then followed by PLP.

Within both fighting and sports it's a well known fact that action beats reaction. If you are reacting to an attack, as the good guys generally are, you are already behind the action curve. Just how behind, scientists have laboured intensely to discover over the last 60 years, and like splitting the atom, they have split the single second into one thousand parts to do it. Breaking down the second helps to clearly define reaction times, the use of millisecond was first used by L Milli in (1922) as 'one thousandth of a second'.

So what did Hick prove and what was the benefit to human movement? Basically Hick experimented with reaction time and the decisions that occur during this process. To be very accurate his research centreed on Choice Reaction Time ("CRT") and it's the "choice" which has been conveniently dropped from most of the writing surrounding this law, which according to Hick slows down as the decision variables increase. In other words, there is an increase in choice reaction time with the logarithm of set size, or put another way, the more choices you have, the longer it takes to choose. There are some statistics around that state; it takes 58% more time to choose between two choices.

That's a staggering amount of time when real time life and death decisions are needed right? HL explores the interference that occurs during retrieval from declarative memory, it also goes on to state that there are occasional savings in response time due to stimulus response repetitions. In chapter one, I explored the different types of memory that humans have and how these relate to stress and the brain. Just looking at the words being used here will give a clue as to what is going on, 'choose' and 'stimulus response' are two examples that are key to understanding the implications of this Law when applied to behavioural based methods of self defence. The message that is relatively clear here is that there is a significant change in data, with practice and stimulus response repetition.

Remember, that a stimulus that brings forth an episodic memory will also bring with it the ability for the mind to pay more detailed attention to that particular thought. Episodic memories are those that are encoded into the mind through an emotional experience. These experiences are capable of coding in the time, place, feelings and details of the event, they are far more real to the mind than attempting to memorise an event to which you are just a passive observer, a little like being told to choose between pressing the left or right buttons when a light comes onto the screen. It's a little dull and there is no reason why, other than your willingness to engage in the experiment to pay attention. Semantic memory is generally concerned with knowledge of the world that we live in, there is a difference between knowledge that is factual and personal experiences that have encoded knowledge and understanding with a greater grounding and meaning.

Both semantic and episodic memory deals with long term, rather than short-term memory. A key difference is that episodic memories encode the actual acquisition experience and the context in which the memory occurred. For any combative or martial art technique to become efficient and effective, the coding process will need to support the intended action. Techniques will have to become linked to procedural memory. Declarative memory deals with facts and data gained from learning. "declarative memory serves to "chunk" or "bind" together the converging processing outcomes reflecting the learning event, providing a solution to the "binding problem" for memory", Cohen, N. Poldrack, R. Eichenbaum (1975).

The sea is wet and the sun is hot are examples of long-term declarative memories. Procedural memory is concerned with long-term memory including complex motor skills. These skills are first coded into the brain and over time become second nature; you do not have to use a cognitive thought process to access the skills. Playing a musical instrument, driving a car, or combative/martial art techniques, are all examples of procedural memory. There is now new evidence that suggests that these skills are habit driven.

It's important to understand the context in which the original research was conducted and to also get a grip on what is happening when the human brain is being programmed by the type of reactions that it will default to in times of stress.

There has been plenty of research into the area of reaction time; one particular piece was done by Schneider and Anderson (2012). Their research explored past research on Hick's Law and its interpretation in terms of information theory, which they based on the Adaptive Control of Thought Rational. Their model produced a set size (number of stimulus response alternatives) that closely resembles Hick's Law. They also account for changes in the set size (stimulus choice options) effect with practice and they explain the stimulus response repetition effects, which together challenges the information theoretic view of Hick's law.

The original research conducted by Hick was carried out in 1952, he used a computer test, to measure the time it took to decide between options and came up with the equation $RT = a + b\log_2(n)$. In basic terms his research confirmed that when faced with choices it takes longer to choose and the more choices that you have the longer it takes and it is from this very simple thought process that up to date reality based methods of teaching were born. Are we humans so very simple? Is the way the human brain works so simple? Does it take a long, slow, encumbering amount of time to make decisions that could put life at risk for example? For some, the answer is a resounding YES and as a consequence they misinterpret this information or worse still, do not have the knowledge that allows for an intelligent exploration of human behaviour.

Research by Schneider and Anderson (2012) found that when practice was allowed the slope of Hick's Law tends to decrease as the number of trials increase, the reaction time decreases, which in turn allows for more choices. There have also been mathematical calculations done that estimate that after about one million trials the CRT will be independent of any set size. So there it is, one million repetitions and your reaction time will be down to zero! Let's remind ourselves what Hick found. Using CRT experiments, response was proportional to log(N), where N is the number of different possible stimuli. In other words, reaction time rises with N, but once N gets large, reaction time no longer increases so much as when N was small, as the number of stimuli rise so the RT decreases.

Kosinski (2010) created a literature review on reaction time. Within the review he discussed practice and errors and what he found would at first seem to support Hick's Law in that, when participants were new to a choice reaction test, they were predictably slower. Once they had time to practice, the reaction times increased. Again very predictable, and to most a logical progression. The results also found that when errors were made, Reaction Time slowed, they also noticed that practice time stabilised the reaction time for up to three weeks. If a system was teaching a limited amount of moves, it would certainly see results based on these facts as the practice that was repeated would have embedded itself for a reasonable amount of time and if further practice was undertaken then the results would bounce themselves on for another period of time. There is no distinction here with complicated routines, if volitional practice occurred, reactions and movements would soon start to get faster with less mistakes.

Now here is the real important part, Stimulus Response and Hick's Law! What Schneider and Anderson (2012) also found is that the graphical slope of Hick's Law can be close to zero for highly compatible stimulus response combinations. The type of responses that were researched covered vocal and manual responses to manipulated stimulus types. Without going into the detail, the explanation given for the close to zero stimulus responses combinations were that the actions required were highly compatible, easy to remember natural movements and that much more pre-experimental practice had occurred prior to test. As a control less compatible combinations were also tested (Brainard et al., (1962); Davis et al., (1961); Fits and Posner, (1967); Longstreth et al., (1985); Teichner and Krebs, (1974); see Morin, Konick, Troxell, and McPherson, (1965) cited by Schneider and Anderson (2012). This evidence supports the age old adage of "practice makes perfect" or a more up to date term might be, "perfect practice done slow and accurately programs the brain to respond fast!"

In the above tests the stimulus responses were chosen for their compatibility with natural behaviour. However, the real point is that it's not a good idea to take what seems to be a logical statement, warp it out of all context and then sell it as the answer to all the problems.

If the above was not enough evidence, there are some that take this rule of CRT and expand it to use a doubling rule. In citing this rule they believe that every decision over and above your first choice will double the time taken to react. A simple piece of mathematics will help us here. Choosing between two choices takes approximately 300 milliseconds (ms), add another choice and we get 600ms, (double the time) another 900ms, another 1 second, 200ms etc - you get the concept I'm sure. What we have is 1.5 seconds to choose between 6 choices, if this were the case, then not only would we see a fantastic staggering effect when it comes to most highly skilled sports like motor racing, MMA, tennis, football, the list can go on and on, we would also in all probability not be the dominant animal on the planet today, as those 1.5 seconds to make a choice between 6 strategies and actions would have made us food rather than the hunter. I know that all the evidence above seems logical, however there still seems to be a belief that there is somehow a difference between combat/martial moves and everyday actions like driving a car and that somehow they are not related! Well that's the problem!, with belief you ignore all the facts that are in plain sight for all to see.

Power Law of Practice

After Hick's Law came the Power Law of Practice ("PLP"). In 1980, Newell, Allen and Rosenbloom published a paper that explored the subject of practice and the performance improvements that it creates along with the supporting mechanisms that allow the improvements to become embedded in the behaviour of the individual. This research considered the chunking theory of learning as a means to explain some of the outcomes of performance that relies on practice. They wanted to confirm the empirical reality that this law was applicable to learning in general, that it was ubiquitous across all forms of learning, rather than just being restricted to skill. The PLP is usually associated with perceptual motor skills. Before I move on with their research it's important to understand a little more about the processes involved in learning skills.

The development of perceptual-motor skills begins early in childhood and continues throughout life, providing that the adult individual continues to expand their skill set. There are three stages to this process of development.

1. Cognitive
2. Associative
3. Autonomous

The first stage looks at what is needed to perform a move or task. This stage requires a certain understanding of the action that is to be learned.

At the second stage, practice is required, another term for this could be "training", where an individual trains a move or sequence of moves over and over again.

The final stage is embedding the moves into the subconscious so that they can be performed without having to pay attention to any procedures that need to occur. The aim here is to produce speed and accuracy, anything other than this would revert itself back to stage two.

Any hand eye coordinated movements fall into the category perceptual motor skills, other examples would be body movement and control, which includes bilateral movement, postural formation and control, auditory language skills, visual auditory skills and any martial based activity. Before any of the higher skill levels can be achieved or worked on, an infant must first acquire the basics, which include rolling, crawling, standing, walking, running and so on until they have good overall control of their body. Once this has been achieved, more advanced skills can emerge, such as running and jumping, catching and writing, these all involve motor skill practice.

The next explanation needs to focus on the perceptual side of this equation. Perception is harder to define, as it's the knowing of how to do something rather than the performance of the skill. Perception skill also has to be separated from intellectual skills, these are generally skills that can be written and defined to allow others to follow the instructions and gain an understanding of how a particular skill is performed. For example, a person could, after some explanation, write a manual on how to play chess. Now imagine trying to write a manual on how to ride a bike, the general principles could be written down, but the 'how' could not. It's the performance of the 'how' part that relates to perceptual motor skills which cannot be gained by simply reading a description of the act.

Once these types of skills are internalized they become part of natural behaviour, in other words the skill becomes an ability, which is performed spontaneously without input from the conscious mind and it's these highly developed perceptual motor skills that can be learnt and developed with enough volitional practice. Here we can see the link between the PLP and the perceptual motor skill ability as, over extended periods of time, the ability is learned and transferred from a simple motor skill into a perceptual motor skill.

The transference occurs and performance speed increases when practice becomes a habit and not just something that is trained a few times a week and that's the biggest difference, if an individual is practicing as a result of habitual processes then the behaviour will soon become ingrained, becoming a perceptual motor skill.

Most would agree that this level of ability is the goal of what they teach, their aim is inbuilt hard wired responses to violent encounters and the way to get to this highly sought after ability, is to train basic moves. Well I think that we are beginning to get the picture that even the habit of walking is a highly skilled ability, it's just been so thoroughly programmed that it's now a perceptual motor skill and one that we find hard to re-program.

The research conducted by Newell, Allen and Rosenbloom (1980) into the ubiquity of the Power Law of Practice theory did not fit the simple power law. They concluded that there were systematic shape deviations in the log-log space, in their words " There exists a ubiquitous quantitative law of practice, it appears to follow a power law. That is plotting the logarithm of the time to perform a task against the logarithm of die trial number always yields a straight line, more or less. We will refer to this law variously as the Log-Log Linear Learning Law or the Power Law of Practice". To summarize, their research found that the law holds for performance measured as the time to achieve a fixed task.

They looked at three learning curves; exponential, hyperbolic and power law. They found that there was a mechanism that was slowing down the rate of learning and that errors in practice decreased with practice and accuracy increased with practice. This was true for different types of learning, which included perceptual motor skills, perception, motor behaviour, memory and complex routines.

This provides evidence that simple basic responses like those that were tested in Hick's Law, will, along with complex movements, all fall into the category of PLP. It is therefore a mistake to focus on simple movements to the exclusion of complex ones as both have the same learning capacity according to the Power Law of Practice.

What is evident from the above is that humans have a capacity to learn complex movements and have protracted capability to remember data. This will help to explain the complicated skills that are involved in sports that have complicated routines like playing tennis, boxing, self defence systems, or actions like typing, playing chess which all involve the ability to learn, memorize, practice and over time internalize so that the activity becomes a part of the perceptual motor skill, no longer requiring complex thought processes to maintain the behaviour.

According to Newell and Rosenbloom (1980) a long standing view is that learning consists of transferring a deliberate, conscious and resource limited process into an automatic, unconscious and resource independent one. What this statement relates to is any process that involves learning, this could be a simple motor skill like writing or a more complex routine of movements like dance, each would have a cost in resources to execution of the task. The cost may involve mental or physical energy, add to this time to execute and practice and we have a process that enables the acquisition of skill. The aim is to be able to reproduce this skill with less time and resources being used, so that the cost in energy is greatly reduced, once this decrease in resources to skill occurs we have a resulting skill which takes less time to complete, less energy is consumed and fewer mistakes occur.

Reaction Time Responses

Let's take a look at some more up to date evidence that relates to these reaction time responses, research by Silva, Cid, Ferreira and Marques (2011) into the attention and reaction times in Shotokan Athletes produced some interesting results. The aim of their study was to analyze the attention capacity and reaction time in Portuguese Karate Shotokan athletes. The participants were physically characterized into weight, height, body mass index and body fat mass percentage and evaluated on Simple Reaction Time (SRT), Choice Reaction Time (CRT), Decision Time (DT) and Distributed Attention (DA).

What they found was that both female and male participants, when tested for SRT, reacted near to the 300 ms mark and that there was no significant difference between the two gender groups. However both the CRT and the DT indicated a significant difference, which was higher in the Dan and 35+ year group than in any other group. The Dan 35+ group also showed a lower percentage of mistakes. The athletes who had more years of practice and were higher in grade needed more time to react to the stimulus than the younger less qualified individuals, however they made far fewer mistakes in their choices than the other group.

Reaction times have been the subject of study for many years, they were first studied by Donders (1868), the results that were obtained showed that a simple reaction time is shorter than a recognition reaction time, and that the choice reaction time is longest of all and it's this CRT that Hick studied.

A study conducted by Etnyre and Kinugasa (2002) cited by Kosinski (2010) found that participants that were asked to react to an auditory stimuli by extending their leg had a faster reaction time if they performed a 3 second isometric contraction of the leg muscle prior to the stimulus. They also found that the pre-contraction part of the reaction time was also shorter, they commented that it was as if the isometric contraction allowed the brain to work faster, or it could have been that this was akin to anticipating the stimuli. Other research on reaction times of participants who were asked to extend either leg in a choice reaction time test indicates that again muscular tension prior to the reaction allowed the limb to work faster. What is evident here is that being ready and preparing the limb with a degree of tension allowed for a faster response.

VaezMousavi et al. (2009) cited by Kosinski (2010) measured arousal in a continuous performance task by skin conductance, and found that while some subjects showed the traditional pattern, others showed the opposite trend. In general, reaction time tended to improve as arousal increased. Anxiety is not necessarily bad. The subject of performance under stress, anxiety and arousal are explored in the following chapter, sufficed to say that in the context of choice reaction times there are procedures that can speed up the time it takes to respond to an expected stimulus.

This brings me all the way back to those that blindly quote a small part of Hick's Law to justify their simplistic approach to human movement and reaction times, knowing how the human body works and how psychology has helped to explain very complex abilities within the brain enables a logical system to be built. One that allows for the complex ability of the human brain and the highly coordinated ability of the body to move in space and time. I have not touched too much on attention, fear or startle reactions that can, in the right circumstances, and with the proper training, increase the body's reaction speed, let alone symmetry or arousal based reactions. So it's fair to say that we have come a long way since the early tests of Hick and certainly Ockham in the 14th century.

Ultimately, simplicity will always be a part of any system, but it does not have to stop there, correct training on stimulus based reactions will get results, scenario based systems will get results, simple techniques, will get results, what matters is how they are trained and what mental processes are engaged in the practice.

Body Reflex

I now come to the main subject of this chapter 'Startle Reflex' ("SR"), before I explore this subject in more depth it's again important that we realize exactly how the reflex occurs before looking at the benefits this understanding can have in a combative/martial application and how it relates to the above material on reaction times. One of the key points has to do with understanding the difference between a reflex and a choice reaction.

I want to look at some of the physiology that occurs as well as the psychology. By physiology I mean bodily reactions that can be predicted. There are two basic categories of reflex that occur within the human body, one is designed as a protection mechanism against a perceived danger or pain and the second is an internal control system responsible for maintaining the homeostasis balance within the body.

The SR is one of the body's first physiological responses to a surprise stimulus, an SR occurs when one of the body's perception senses is stimulated by an unanticipated incoming stimulus, the body reflexes involuntarily as a response to this external stimulus. We know from research and experiments carried out that the body has this in-built protection system designed to protect the body from harm. The SR is also referred to as a "Flinch". All humans will respond in a certain manner when a startling stimulus occurs, for example; blinking, upward movement of the shoulders, head tucking in and down, bending of the arms and their withdrawal into the core, as well as bending of the legs and their withdrawal into the core. There are several "facial expressions" and "various twisting of the whole body and limbs" Chapél, (1991). In the majority of Martial Arts schools there is no inclusion of this within the training of self-defence techniques. Techniques usually start from a punch, grab or kick scenario.

So let's recap for a moment, reaction time is the time that it takes the body to respond to a perceived stimulus from the environment. Reactions can be categorised as Simple Reaction Time (SRT), Choice Reaction Time (CRT), Decision Time (DT) and Distributed Attention (DA). We then have reflexs, while these may seem similar to begin with, they are far from similar and indeed they are very different. There are a number of ways in which reflexes can be categorised, some reflexes are required to control the body's homeostatic state, for example a baroreceptor reflex causes a decrease in heart rate, that follows a distention of the carotid sinus. Also there are somato-visceral reflexes, that cause either a vasoconstriction or vasodilation of blood vessels that result in cooling or heating the body, these have also been found to be a result of high stress situations and the adrenal dump. Somatosomatic reflexes are primarily associated with body movement.

Reflexes are unconditioned involuntary responses and are the body's mechanisms for protecting itself from perceived danger, they are also faster in their action time, than a 'reaction' of any kind. Reflexes are usually a negative feedback loop and act to help return the body to its normal functioning stability, or homeostasis. A simple example of a reflex is the blink response to either an incoming object towards the eyes or an auditory noise

A type of reflex associated with the somatosomatic category is the stretch reflex, this is also know as a monosynaptic ipsilateral segmental reflex. This action involves the contraction of a muscle that is reacting to a stimulus either within the tendon or the muscle itself, the stretch reflex is fast and has a fast onset and offset. It is commonly found when the body is having to rebalance itself, due to the effects of gravity, keeping the limbs extended so that an individual does not fall. The classic example of this type of reflex is the patella reflex. This reflex is a stretch reflex and is initiated by tapping the tendon below the patella, or kneecap. The same reflex action can be found in the Achilles reflex and the biceps reflex.

Flexor reflex, this is also know as the withdrawal reflex and is the one that is most relevant to the combative/martial instructor, the actions involved can cascade to other limbs and can also, depending on the strength of the stimulus, effect the whole body. This type of reflex is a protective polysynaptic ipsilateral segmental reflex, pain is the main stimulus that triggers the contraction of ipsilateral flexor muscles and the extensors to relax in the withdrawing limb, at the same time the opposite reaction occurs in the other limb, the flexors relax and the extensors contract, which enable the cerebellum to control equilibrium in balance and posture of the whole body.

An example would be when an individual steps on a sharp object, the leg that has received the pain stimuli pulls away from the source of the pain, while the other limb takes the full weight of the body. This type of bilateral reflex is also called a crossed extensor reflex, indicating that the reflex is contralateral and occurs on the opposite side of the body that received the pain stimulus.

In a paper by Mann (2011) he explains an understanding of the crossed extensor reflex and how it supports the continued balance of the body and how it differs from the lower strength stretch reflex. Usually, the postural muscles experience relatively slow, sustained stretches and the anti-gravity muscles, of which the quadriceps is an example, are pulled upon by gravitational forces. This steady force sets up a sustained discharge in each group of afferent muscle fibers, however the discharges in different fibers are not synchronized as they are when the tendon is tapped.

In addition, longer, larger stretches are able to excite secondary muscle spindle receptors which also have connections with homonymous alpha motoneurons, di and trisynaptic ones. These longer, larger stretches therefore activate the alpha-motoneurons by both monosynaptic and polysynaptic reflex pathways. The resulting reflex contraction of the muscle is called the stretch reflex. The polysynaptic effects are not seen in the tendon tap reflex for two reasons: (1) the brief stretch does not excite secondary spindle receptors and (2) the brief input over the polysynaptic pathways arrives after the monosynaptic input and finds the alpha-motoneurons in their refractory periods and therefore cannot cause them to discharge again.

In controlling posture, the asynchronous discharge in mono and polysynaptic pathways induced by gravitational forces on muscles, comes in the alpha-motoneuron with other activity from within the Central Nervous System to produce a contraction that just balances the gravitational force. If an additional force is applied, stretching the muscle, additional tension is developed by the stretch reflex to counteract that force. In this way, the stretch reflex serves as a mechanism for maintaining an upright body orientation under a variety of load conditions; the mechanism is automatic ("unconscious") and fast (19-24 msec for the quadriceps in man).

Here we have the first example of the time that it takes a major muscle group such as the quadriceps to respond to a reflexive stimulus, what is very evident is that the 19 -24 msec time span is greatly different than the 300 ms mean CRT that was discussed earlier, this indicates the speed at which the sensory motor neurons engage with the motor neurons and fire the muscle to move, all this is executed without conscious lag.

So far what we have found is that the protective reflex mechanism (withdrawal reflex) is adjusted due to the strength of the stimulus and that a higher strength cascading reflex can cause more than one limb to move and that usually the body tries to engage the anti gravitational muscles of the opposite limb to keep equilibrium and balance. The cascading reflex that is due to a higher strength stimulus is known as 'irradiation'', it can be predicted that a certain modeled behaviour will occur depending upon the site that receives the stimulation, this is often called a local initiation site.

What this essentially achieves is bilateral reflex control, which enables the body to protect itself and maintain balance providing a means of further movement. If the stimulus reaction to pain is high and the site for the stimulus changes to the groin area for example, then both sides will react via the crossed flexor reflex and with a very high level strength stimulus the reflex will cascade to the upper limbs and even as far as the head, this very intense stimulus to the groin results in all four limbs and the head being retracted in towards the core of the body. The body then collapses to the ground as can be witnessed when a strike to the groin area occurs.

What Should you Teach?

Firstly if you teach reality-based self-defence techniques and exclude this from your curriculum, you are not giving your students all the knowledge possible to enable them to protect themselves. You must however remember that the startle reflex only occurs when you are startled, this may be an obvious statement, but it is very important that a clear distinction is made between: being caught off guard and completely by surprise with no awareness of the impending attack; and being attacked and responding in some kind of trained manner to a confrontation that has already begun, a preamble or a pre violence dance has occurred.

An involuntary startle is due to stimulus being perceived via your eyes, ears or touch sensory system. These can be categorised as Auditory Startle, Visual Startle and Sanatoma Sensory Startle. The first thing to understand here is that this reflex cannot be trained out! When an individual is continually exposed to a particular type of stimulus the body experiences an increase in our body's reaction to startles, this process is called "sensitization" where as a decrease is called "habituation". This means that the body will habituate to a certain point when it continues to receive startle stimulus.

For example in the film "We are Solders" with Mel Gibson, when the journalist first appears on the battleground, the explosions startle him. However, during the end scene when others turn up after the battle, his startle reflex had been habituated to explosions, where as the new journalists all startle. After a period of time the body will return to normal reactions to this stimulus and the habituated response will become extinct. What this tells us is that we can habituate being caught off guard when attacked and to a degree we can train down a startle reflex. However we would have to be continually training in a method that created a startle all of the time, as soon as we suspend this type of training the reflex will re-initiate itself.

When the body enters a startle reflex it moves in a manner that is faster than any other type of body movement, remember the 19 – 24 ms reaction time of the stretch reflex above as opposed to the 300 ms for a CRT. It simply cannot be reproduced by any conscious thought process. Any idea that you can train in a way that uses a flinch or startle response as part of your initial conscious thought process to respond to an attack is unrealistic, due to the physiology of the body.

Spinal Reflex

When a stimulus is received via the proprioceptors that are situated within the skeletal muscles, tendons, skin and joints, sensory neurons are transmitted into the spinal cord, here they interact with spinal motor neurons, which then bounce back to the body part, initiating the reflexive movement. If the reflex action requires a greater response from several muscle groups, then the interspinal tract of the spinal cord gets involved.

"The resulting reflex is a flexor reflex, that protects the limb or body part, by initiating a withdrawal movement away from the stimulus ' this is different from the stretch reflex where the response occurs in only one muscle via a monosynaptic pathway" Tyldesley and Grieve (1996). They also go on to describe how inter neurons carry impulses across to the opposite side of the spinal cord in the 'crossed extensor reflex' resulting in the activation of the extensor muscles of the opposite limb to prevent the body from falling over when one leg is flexed.

The spinal cord is the first and last stop for these types of reflex actions, which are evident in new born babies and shows itself with several different types of reflex actions, some of which are lost as the newborn grows. As the whole nervous system develops more complex actions are programmed into the reaction and movement control centres, that will ultimately allow the infant to stand, walk and feed itself. However one action will always remain the responsibility of the spinal cord and that is the processing of the spinal reflexes. What this understanding provides is that there are different levels of motor control of the human body, these are, spinal control, brain stem control and higher cortical control. According to Tyldesley and Grieve the neuro activity spreads to the upper limbs, the arm on the same side extends and the opposite side flexes. This pattern of movement seen in spinal reflex actions form the background to normal movement. This relates itself directly to the chapter on bilateral symmetry and provides evidence that the body has pre-programmed response actions that are innately hard wired into our brains and it's this type of movement that allows the human body to move with speed and power.

Brain Stem Control

The control centre within the brain stem has a certain control over the spinal level in that it is responsible for maintaining posture and balance control of the body. It is also the control and activation centre that receives input from the proprioceptors within the cerebellum, it receives neurotransmissions from the eyes and the ears as they also perceive incoming stimuli from the external environment. This is where the Auditory Startle and the Visual Startle are processed. The blink reflex is a reflex which is designed to naturally protect the eyes and is present from the time that infants first opens their eyes. Multiple cranial nerves are involved in the process of blinking, as is the brainstem, the blink is often over before someone is consciously aware that there was a risk.

Several different stimuli can trigger the blink reflex. Anything which touches the cornea will cause someone to blink, blink reflexes usually occur when objects appear to be on a direct collision course with the eye, as for example when something is thrown in the direction of someone's head. Very bright light also stimulates the blink reflex, as do loud noises. In conjunction with the blink the whole head may also move away from the same stimuli that causes the blink reflex.

One reflex that although interesting, is not required for movement other than an internal reaction, is the reaction time to unpleasant or spoiled food, the stimuli for this would first be perceived via the olfactory system. Boesveldt *et al.* (2010) cited by Kosinski (2010) noted that unpleasant odours (such as from spoiled food) might have great relevance to survival and health. They found that reaction time to unpleasant food odours was faster and more accurate than reaction to pleasant odours and to non-food odours.

Higher Cortical Control

The motor system of the human body is controlled by the cortical motor areas, basil ganglia and cerebellum. They are responsible for the activation of movement while attending to the constant input from the external sensory environment. This is the processing centre that allows for conscious control over movement and also allows for the interruption of movement already planned.

Another factor to be considered with regard to being startled and the severity of the startle is the situation and environment that you are in at the time. If you are in a dark alley and are alone at night, then the intensity of the startle may well be greater than the same alley during daylight hours. The key areas to consider with regard to Startle Reflex and combative/martial arts are; can we train a response that can be used? Can we move intentionally at a speed approaching "SR" speed?

There are recorded accounts of people who have trained intensively reacting to a startle stimulus in a trained response manner while under extreme emotionally charged situations. This would indicate that although the reflex cannot be trained out, it could be substituted for movements similar to self-protection moves that you have trained for. I am not talking about full on blocks or attacks. I am talking about shielding moves of reflex hand swipes across the face i.e. programming a response that will help protect you if surprised.

With regard to moving at the same speed, we have to understand the physiology behind the reflex. The neurons that fire during the reflex action, never reach the conscious parts of our brain. The body has to switch off all the prime mover fixator muscles and instead use the fast twitch muscles, known as our Antagonistic muscles. Typically, the empirical evidence indicates that the body parts that are moving during this action first move back towards our core.

The head retracts, shoulders hunch, arms bend and retract, knees bend and our legs withdraw to our core, briefly wanting to return to the fetal position. Knowing this and understanding the body's natural reactions will allow instructors to prepare students for a stimulus based trained response to a surprise attack. What comes next is the dump of a chemical cocktail into the blood stream to enable the body to cope with the impending violence. At this stage we will also enter a state of mind that will either help or impede our survival.

It's also important to remember at this point that the attack is a surprise! Your attacker could be lying in wait for you, or stalking you ready to attack at a moment of their choosing, when they perceive you to be at your weakest. There is unlikely to be any verbal warning that the attack is coming, therefore coping strategies for a verbal encounter should not be much of a concern with this method of attack.

This is very much about prior knowledge of your body's natural protection mechanisms and the simple fact that your only chance of a response during this surprise attack is to train a stimulus based, programmed response. There is a distinct chance that you may even be shocked into a freeze state, one in which you are incapable of any response. Now we are entering the realm of our body's physiology and the adrenal dump. What this will do is send the body into a high state of emotion, knowing what this feels like and understanding it, is the first step to coping with the effects on the body. To clarify, this is a surprise attack! First we enter a startle reflex and then the body goes into some type of fight or flight response, due to the adrenal dump. When the situation is changed to a perceived encounter first, then the adrenal dump will come first.

Early in this chapter I spoke about the need to have a very limited response against a surprise attack. We need to look at this in a little more detail here. How are we going to be surprised? As a combatant/martial arts instructor, one of the first things that should be taught is awareness, awareness of your environment, the potential dangers and how to avoid them. Let's face it, with today's technology how many times do you see people walking down a street with ear phones in and music blasting out, or they may be totally engrossed in a phone conversation. Colour coding awareness levels has been put forward before (Cooper 1989), with awareness levels running from white to black, white being totally unaware and black being in the middle of combat. It's simply not possible to be totally aware at all times, we are all capable of being surprised. Even if we are expecting an assault we can still be startled. So what type of shielding moves of reflex hand swipes should we train.

As far as possible they need to mimic the movements that would be made during a startle reflex, it's no good trying to program in something that is far removed from the actual moves. We know that both hands will work in a symmetrical manner, this means that they may both retract together, depending on the strength of the stimuli. This then can be used in our favour, bringing both hands in and back and up to cover the face would be one example of a shielding movement. Another could be just one hand being swiped across the face, as if trying to swat a fly away. Both have to be programmed into the responses.

We must remember here that these moves are only our initial reflex responses to being surprised; we have yet to respond in a significant manner. As our reflexes are part of our body's autonomic nervous system there is not a great deal that we can do, other than train it into an habituated state that will need constant re-enforcing to prevent extinction. An area of training that I also mentioned is stimulus training as opposed to scenario-based training. This is where training a response to a stimulus takes priority over knowing what the attack will be, what's key here, is how this type of training can be started and then progressed so that true un-known attacks can be handled effectively by any student.

There are reflexes involving the heart. Baroreceptors (pressure receptors) in the carotid sinus, in the aortic arch, increase their discharge rates in response to elevated blood pressure. This increase in activity leads reflexly to a decrease in sympathetic discharge, resulting in peripheral vasodilatation and an increase in parasympathetic discharge primarily through the vagus nerve, resulting in decreased heart rate. The net result is a decrease in blood pressure, which then triggers the brain to instantly react by shutting down consciousness. This reflex has been known for many hundreds of years within the martial community, one target that takes advantage of this reflex is a hand-sword to the carotid artery.

One thing that we must always remember is that as members of the homeo-sapien species we are all susceptible to certain reflexes that can cause involuntary responses, just like the crossed extensor reflex above, there are others that will automatically create greater problems while engaged in physical encounters

The aim of this chapter was to provide evidence to base your training on and not to just use limited parts of knowledge to restrict the type of training that you use. It is very important that the training that you immerse yourself in is based upon solid up to date knowledge, remember that your life may one day depend upon what you have programmed into your brain. If you at least understand the reasons why you do certain moves, you will then be better placed to pass this knowledge onto future students that may cross your path.

9 PERFORMANCE UNDER STRESS

Walter Cannon first coined the term "fight or flight" in 1915, since then it has become linked with combat arts, door supervisors and just about every martial artist. Most teach an understanding of the adrenal dump and the effects that it has on the human body when in a fearful or stressful situation.

Being stressed and in fear are two psychological and physiological responses. Stress is a term that is hard to explain and can occur in varying degrees of intensity. It is linked to homeostasis, this is a term used to explain a system that is self-regulating and the human body is one such system. It has controls and mechanisms in place to maintain the ideal operating environment within. This ensures that a stable equilibrium is maintained at all time.

Our bodies have several systems that all interlink to maintain a healthy organism. Such as blood pressure, temperature or oxygenation. At a micro level, blood itself is monitored and adjusted and contains dissipated proteins, glucose, mineral ions, hormones and carbon dioxide. Anything that affects the balance of this system could cause stress. For example, you find yourself getting hot and are unable to remove clothing, your body temperature spikes and you start to feel stressed, not fear.

Fear is generated by a stimulus, and in some cases a predisposed state of mind that is a result of behavioural tendencies. What do humans fear? is it always something that can harm the individual, such as a snake or shark. There are those that fear natural occurrences, darkness, walking down the street, these do not at first present an immediate threat to the individual. Interestingly some humans that live in remote destinations that have never even seen a snake will recoil from one when first seen, indicating that this is an evolutionary mechanism.

Stress – Anxiety - Fear!

An individual experiencing fear would usually be as a result of being in immediate danger of attack or death. The emotional fear response leads to two basic choices, stand and fight, or turn and run. It is at this point when the stimulus causing the fear could be so intense, that it triggers several different responses. Before I delve into an exploration of fear, let's first explore "stress", this is a person's response to stimuli received from within the mind. The body activates the sympathetic nervous system sending signals to the body to react in a certain way. The reaction to stress unbalances the homeostasis equilibrium within the body. Behavioural changes can occur when stress takes a hold of the individual.

Stress is mechanically different from fear, yet it is not different from anxiety, fear and anxiety are closely related, separating the difference between them is more a matter of semantics than a real difference. However fear is a stimulus event, an emotion, whereas anxiety is a post stimulus event. To feel anxiety you must have first felt fear a mammalian mechanism for protecting the organism. Anxiety is an uncertain feeling within oneself. Anxiety occurs in situations only perceived as uncontrollable or unavoidable after an initial brush with fear and as such can have a significant effect on performance. It is therefore easy to consider stress as another way of describing anxiety. Anxiety can be categorised as either state or trait anxiety "state anxiety is a momentary experience of anxiety. Trait anxiety is a predisposition or proneness to be anxious" Fernandez-Ballesteros. (2002).

Anxiety can create different physical effects within the body and are not exclusive to any one individual, they may include heart palpitations, muscle weakness, tremors or tension, fatigue, chest pain, hyperventilation, loss of mental processing ability, these can all happen in various degrees of severity. External signs of anxiety may include pallor skin, sweating, trembling, and papillary dilation. Anxiety can also affect people in different ways, as some individuals are more prone to experiencing these feeling than others. Individuals that experience anxiety on a regular basis are diagnosed as suffering from an anxiety disorder, such as PTSD or a panic disorder. Other anxieties are focused on external stimuli that are perceived as being dangerous or threatening such as animals or physical danger and fall under the category of 'phobia', social phobia, agoraphobia or claustrophobia are examples of these. Understanding if a person has a trait propensity allows for specialised training, focusing on habitualising these feelings. The techniques for dealing with anxiety, especially when related to performance, are varied and need to be understood to ensure that an individual's performance is at its peak especially when dealing with a violent situation or deadly force encounter.

Asken (2010) discusses performance failure at a critical moment, when an individual experiences anxiety during a high state of emotional arousal due to confrontation or deadly force encounter. He refers to this failure as "choking", associating the word with being unable to get through a situation, rather than the lack of air arriving at the brain. Choking could also be a response to a high state of fear or a culmination of both anxiety and fear. From a trainer's point of view, understanding the character or traits that an individual has are invaluable "those who fare well in high stress situations (don't choke) are low on hypnotic ability/absorption – a measure of hypnotic ability; low on neuroticism – the tendency to fixate on catastrophising thoughts and negative emotions; and high on repressive coping – a left brain hemispheric localized behavioural tendency that has been shown to inhibit the inter-hemispheric transfer of negative effect (emotions) between hemispheres of the brain" Carlstedt (2004) cited by Asken (2010).

What this gives evidence to, is the fact that traits of anxiety do not occur in some individuals. What have these individuals with a high coping strategy got that others with low coping strategies don't have? Is it behavioural tendencies that have been experienced over a lifetime? It is perhaps understandable that an individual with constant exposure to significant caregivers who themselves portray or have anxiety problems, would inherit these traits. In effect, they create episodic memories that are encoded early into their children and so over the years develop a tendency towards anxiety traits, rather than experiencing a brief state of anxiety "there are physiological, cognitive and behavioural components to anxiety" Fernandez-Ballesteros. (2002).

These components all help to explain the category of anxiety, which in turn affects the performance of an individual. Behaviourism is a well known school of thought first introduced by J. B Watson and is a subject that we will need to look at if our aim is to fully understanding why some do and some don't perform well while experiencing stress. There are psychological techniques and strategies for dealing with the effects of anxiety, these also have a significant role to play in fear management as well. However, any technique needs to have a base on which to build the strategies. Teaching an individual the coping strategies is one part of the equation; the other is the science that explains the processes that are happening within the brain and mind of the individual in question.

When I first introduced this chapter I talked briefly about homeostasis, the system that controls and monitors the human body, always maintaining equilibrium between the systems that are required for a healthy individual. However, we all know from experience that maintaining a controlled state of mind and body very rarely occurs.

We are all exposed to incoming stimuli that can have a devastating effect on our minds and our bodies. Not only do we have the issues of stress and anxiety to deal with, but we also have fear. For the most part, the majority of humans no longer have to experience the traumatic effects of primeval man, when we relied heavily upon our instincts to survive. We live in relative safety and comfort, with the minority being exposed to high states of arousal due to confrontation or deadly force encounters. However, what happens when we are unexpectedly exposed to high stress situations and what about the professional, how do they adapt to new or improved behaviours and learning methods?

I live in a quiet part of the United Kingdom, well away from the major cities of London and Birmingham; we very seldom have to deal with a fearful situation that affects a large group within society. Terrorist attacks like the one that happened in London in July 2005, when three underground trains and a bus were attacked and blown up, or bigger still the twin towers attacks in September 2001, are without a doubt shocking to the world. One can only try to imagine the fear and anxiety that spread like a wildfire in the Australian outback. Then one day in May 2008 an individual walks into a local restaurant where I live and tries to explode a bomb in the toilets, fortunately the only thing he damaged in the process was himself. Suddenly the far off world of terrorist attacks are a little more real than before when they were happening some distance away from my everyday life. These events were almost instantly available via news media to the whole country, and in the case of the London and New York attacks, the world.

With global, instant communications, people find themselves affected by these events creating anxiety and fear in a much wider population base than ever before, the closer you are to the actual occurrence, the worse it will be. The human psyche has an ability to distance itself from experiencing traumatic events and the opposite is also true "distance in war is not merely physical. There is also an emotional distance process that plays a vital part in overcoming the resistance to killing. Factors such as cultural distance, moral distance, social distance and mechanical distance are just as effective as physical distance in permitting the killer to deny that he is killing a human being" Grossman (1995).

Here we see distance from the effects of a terrorist attack creating a sense of separation, in both emotional feeling and the physical effects of the event, the closer an individual is to an atrocity like the London bombs than the greater the severity of anxiety and fear felt. This also indicates that a greater trauma will occur the closer an individual is to the epicentre of the event. There will also be a greater likelihood that some will suffer PTSD, this will need to be dealt with to prevent long-term chronic anxiety or depression.

Whenever humans are exposed to situations that cause trauma, the first experience is usually the most vivid and emotional, as long term episodic memories are being encoded into the brain. Once this encoding has happened and the traumatic event has past, any further exposure to the same stimulus will cause a re-uptake of the stimulus into the brain, the causal effects of this will be a return of the feelings and stress associated with the stimulus, you will now have a post trauma event that creates stress, add the disorder label and we have PTSD. This could happen with any everyday event or an evolutionary mechanism designed to protect the human organism, which to the observer could seem utterly un-important. All that has happened is that some individuals may be predisposed to experiences of heightened fear or stress.

For example, one day a child of say 7 is a passenger in their parent's car, during a nasty storm, the winds have increased to above gale force and the trees are beginning to bend, rubbish is being blown onto the road and the parents begin to stress, maybe I should have stayed at home today like the news person recommended, did I really need to go to the store? Then suddenly a tree is blown in front of the car causing the parent to swerve and stop with frightening effects skidding to an abrupt halt! All the time the child is taking in the stimuli from the parent and its environment, fear is building, as they really do not understand what's happening, all they know is that the ones that they trust in are stressed and in fear of their safety, then the tree happens. From that moment on their little finely tuned brain has experienced a trauma and one that has also been encoded with the emotion of how they felt, who they were with and what time of the day it was.

Now a few weeks down the line, they are at school and the wind picks up, they hear the teacher saying that they hope it not as bad as a few weeks ago and that's it, the memory returns, the child begins to stress at the winds speed, the feelings that surround the trauma return, they come crashing back into the forefront of memory, they compound the traumatic event that further leads to hard wire imprinting within the brain and there you have it, a "post traumatic stress event" that is labeled PTSD. Now every time the wind picks up the child returns to that day in their mind and they begin to stress over the wind, the wind becomes their stressor! Usually PTSD is associated with traumatic events that occur during war, more recently professionals that attend accident and emergency scenes have also been diagnosed as suffering from PTSD, just because it's a child does not mean that they can not suffer from the same disorder.

During exposure and dependent upon distance and the stability of the underlying character traits of the individual, there will be individuals that go through a process of change both physically and mentally. The following extract is a good example of the cause and effects of violent anti social attacks on society "the traumatic events of September 2001 have dramatically changed our lives forever. Americans and people around the world were shocked by the unprecedented attacks on civilians. The anthrax scare further shook the complacency of Americans. We must all live with heightened vigilance and increased fear of more unprecedented attacks. With the attacks and their aftermath, many Americans experienced levels of emotion, particularly fear and anxiety, to which they are unaccustomed. Many experienced acute post traumatic symptoms (Schuster et al. 2001; Gala et al., 2002; Schlenger et al., 2002; Silver, et al., 2002). Although some people do not recover well from the trauma and develop chronic anxious or depressive conditions, in the vast majority of cases, these feelings recede, and people go on with their normal lives. These people will not return to their previous state however as the external world has changed, and their internal psychological states have changed too" cited by Schulkin, (2004).

This underlies the importance of understanding that events in life will have an effect on both the psychological and physical state of individuals. Homeostasis is the body's control systems kept in a state of equilibrium, "Allostasis," a term used by Sterling and Eyer in 1988 (McEwen 1998), refers to levels of activity required for the individual to "maintain stability through change" cited by Goldstein, and Koplin (2007), allostasis is a concept that within the human body there are "processes by which mammals maintain viability following physiological change due to adversity" Schulkin (2004). This concept plays a critical role in understanding that through experience, the underlying state of the brain and the control mechanisms of the body change. They do not forever remain constant as in homeostasis state.

Schulkin (2004) explains that, allostasis allows for a change in state whereas homeostasis does not. Allostasis creates set points within the body that fluctuate as a result of experience and in doing so allows for continuous development and fluctuation of new physiological and behavioural set points. The main underlying explanation of allostasis is that of change to maintain a state appropriate to the current and anticipated circumstances. As an individual involved in constant exposure to high emotional states of arousal, allostasis fits well into the understanding that there are causal effects from exposure to anxiety and fear stimulus.

It is these effects that can begin to change the make up of the individual allowing them to build coping strategies "during a fear evoking situation, both central and peripheral processes are highly active in response to threat and in anticipation of further harm. If one survives the danger, the

fear motivates changes in the behaviour in anticipation of encountering a similar danger in the future, A new set point or threshold for activation and elicitation of defensive responses is attained" Schulkin (2004).
This provides evidence that when the professional police officer for example, is constantly exposed to fearful stimuli, habitualisation occurs. To reinforce this learning behaviour and ultimately enhance performance, mindful attention needs to be applied to create episodic and procedural memories. We know that this type of attention can change the synaptic pathways in the brain, leading to enhanced performance and a change in psychological mindset. Experiencing fear and then using these experiences to create a degree of protection from the causal effects of anxiety and fear will also lead to a much more stable and healthy individual in later life.

The region within the brain responsible for producing the activity that leads to the release of hormones, that in turn produce the fight or flight response is the amygdala. The amygdala is an almond shaped mass of nuclei located deep within the temporal lobe of the brain. It is part of the limbic system, which is involved in producing the thoughts of our emotions and motivations, especially those that are involved in an individual's survival. The amygdala is involved in the processing of emotions such as fear, anger and pleasure. The amygdala is also responsible for determining our memories and where the memories are stored in the brain.

Part of the process that happens when the amygdala produces the chemicals responsible for the fight or flight, is that they also trigger an event to encode an episodic memory " the hormone that the amygdala triggers temporarily enhances memory functions, so that the awful experience that triggered the response will be vividly encoded and remembered. Such traumatic memories last and they are potent, long after calm has returned, even years later in some cases they are likely to be recalled with terrifying ease" Gardener (2008).

It is no wonder then that when we experience continued exposure to the memory, as well as the feeling, that we are able to recall the event easily and eventually, over time, a certain amount of habitualisation will occur and at the same time changing the allostasis set point within the mind.

There are cases of human patients with focal bilateral amygdala lesions, due to the rare genetic condition Urbach-Wiethe disease. Such patients fail to exhibit fear related behaviours, leading one to be dubbed the "woman with no fear". This finding reinforces the conclusion that the amygdala "plays a pivotal role in triggering a state of fear" Wikipedia (2013).

Fear is the perception of impending danger of harm or death to an individual and as a consequence to this is one of the main motivators of the body to respond and adapt, leading to survival. Adapting to physical and psychological consequences of fear and at the same time allowing for a new set point to occur from the role of allostasis, will again require us to explore some of the mechanisms that the human body has at its disposal to deal with high emotional states of arousal. Some are maladaptive rather than adaptive.

Internal mechanism responses include; raised heart rate, perspiration, increased breathing rhythm, loss of bowel control, sensory inhibition.

Physical responses and disruptions include loss of body coordination, vision, hearing impairment, death grip, freezing.

Mental responses and disruptions, include increased anxiety, a slow down in mental problem solving, loss of environmental awareness.

Internal Physiological Responses

The physiological responses to incoming stimuli (fear) that are innately recognised as an impending threat to the survival of the organism start within the amygdala. The amygdala triggers a firing of neurons within the hypothalamus. The hypothalamus is located below the thalamus, just above the brain stem. This chain reaction then triggers the pituitary gland to release a secretion of the hormone, Adrenocorticotropic (ACTH).

The adrenal gland releases the neurotransmitter epinephrine, resulting in the production of cortisol, which raises the heart rate creating an increase in blood pressure. This creates a change in the allostasis system controlling blood sugar and suppresses the immune system. All of this is designed to prepare the body for a large use of energy over a short period of time, to either fight or run.

Physiological changes are designed to allow the human organism to increase the overall performance of the human body for short periods of time, increasing speed, strength and heightening other senses. Yet all too often when the heart rate increases beyond normal levels, maladaptive responses also set in degrading performance. We know this to be the case, yet there is also evidence that suggests that the envelope of adrenal responses that occur during the fight or flight response can be increased. What this allows for is an increase in performance rather than a degrading of performance and this is a key area to understand if we are to stand a chance of training individuals to understand what is happening and how to use this understanding to increase performance.

From an evolutionary perspective, the fight or flight response provided prehistoric humans with the mechanisms to rapidly respond to threats against survival. The sympathetic nervous system is the mechanism used by the brain to meet these requirements. What the above provides evidence to is the fact that humans have these responses to enable them to survive for longer and to reproduce. However there are maladaptive responses that can occur during the fight or flight response, these are not necessarily supporting the survival instinct.

Physical Responses and Disruptions

Death Grip

Vestigial traits are evolutionary remnants of features useful in our ancestors and our distant past. These behaviours are now either useless or have been adapted for a different use. One such behaviour is the "grasping reflex" of human infants, which is also referred to as the "death grip" in adults. Placing a finger into the palm of an infant will result in a reflexive action, with the infant immediately and securely grasping it. This grasping reflex is also evident in the feet of babies, they will grasp with the toes in the same manner as the hand. This reflex got its name 'death grip', from the behaviour of swimmers in continuing to hold onto heavy weights while they were being taken to the bottom of the pool.

Interestingly, a relative once reported an event that created a direct link to this grasping reflex. With infants, the reflex is lost relatively early in their developmental life, however a lady of older years was stood outside of her car, with one hand on the steering wheel and one on the ignition, she turned the key, the car being still in gear started to jump forward, rather than release the key and the hand on the steering wheel, she continued to reflexively hold on to both, even after injury started to occur. The car did not stop and she never let go until the car came to a stop by hitting a wall. Evidence that although the grasping reflex was expected to be extinct by this time, a stressor, ie the car moving and catching her by surprise was enough for the grasp reflex to revert back to a mechanical response. This would certainly be termed a maladaptive behaviour if it put the individual at risk of injury, which in this case it did.

Freezing

Another evolutionary vestigial trait or reflex is the action of freezing while engaged in what the individual considers to be a life or death situation. In our distant past, if confronted by a predator that was capable of out running you, a freeze response may well have helped our survival or being overwhelmed by an animal and knowing that there is no hope of survival. To understand the freeze response you have to understand how the two systems of the autonomic nervous system (ANS) function work. One is required to activate and the other to deactivate the survival responses that are innately built in to humans.

The ANS is a network of nerve fiber that extend throughout the body; it is made up of two specific systems. The sympathetic branch activates the fight or flight response. It communicates to the body the responses required for the high stimulus response, raises the heart rate, increases muscle tension, causes the eyes to dilate and the mucous membranes to dry up. All of these responses help an individual's performance; they fight longer and with more determination, run faster, see better and breathe more efficiently. The time taken for this response to activate can be measured in microseconds, less than the amount of time it takes for the heart to beat twice.

The parasympathetic branch activates the return to normal homeostasis balance within the body. For humans, the freeze response usually occurs when we're terrified and believe that there is no chance of escape or survival. It can happen to victims of rape, armed robbery, car accidents or even violent aggressive encounters. It's more likely to occur in un-trained individuals, although even the professional can be at risk if they are not aware of this reaction. Understanding the freeze response will help an individual to cope with this response and mentally plan against it.

Perceptual Narrowing

This is commonly known as "tunnel vision" The incoming visual stimuli, that can usually be alternated between focused and peripheral vision, excludes peripheral in favour of concentrated vision onto one area of focus only, the pupils dilate to help see with increased clarity, you are focusing on the face. This type of response can, in one sense, heighten survival of the individual by allowing a greater amount of detail to be processed, while excluding peripheral stimuli. This is all well and good until a threat emerges from outside the focused tunnel vision zone or, the person in front of you moves their hand to their back pocket ready to deploy a weapon.

Audio Exclusion

This response is very similar to perceptual narrowing however, this time the audio input is focused and rather than taking in background noise, only that which is directly close to the individual will be heard. Relating this to a violent encounter, imagine that you become involved in a face to face encounter with a verbally aggressive guy, pouring out streams of abuse, your adrenal dump has occurred and the only thing you can hear is this guy's noise, you never hear his mate walking up beside you, or the flick of the knife when being deployed! Or worse still, the cocking chamber of a gun from the guy behind you.

Motor Control Interruption

Motor control is another area that can create a maladaptive behaviour. Losing the ability during a fight or flight response, to use both gross and fine motor skills could seriously put your safety at risk. Hand to eye co-ordination requires both gross and fine motor control, losing this would mean not having the ability to point and fire a weapon, defend an incoming attacker that was armed with a blade, or just a very aggressive punch heading straight for your head. After all those years of practice in a safe environment, you find yourself in unfamiliar circumstances, with no prior training and awareness of the possible responses of the human body, and find yourself devoid of control over your limbs.

Internal Psychological Responses

Awareness Failure

Awareness failure also integrates into general awareness of incoming stimuli via the senses of the human body, however the specifics of awareness when related to a shut down due to the fight or flight response are concerned with awareness of our general environment and a lack of ability to be aware of activity that surrounds us. This is also linked to a loss of peripheral vision, shutting down this ability inhibits the mind in paying attention to stimuli outside of the focused point.

Decision Errors

While within the mode of fight or flight, your ability to make critical and, what would seem usually easy, logical decisions can be impaired, slowing down reaction time and leading to degrading of performance.

It is very important at this point that we understand the physiological differences that are a result of physically induced high heart rates and fear responses. Both create sky rocketing heart rates, there are mechanisms that alter the homeostasis state within the body and these are different for both. The controlling system is known as thermoregulation and is an important aspect of human homeostasis. . The responses that the body creates when it is physically pushed to extremes are consequences of our mental processes, to some extent they are wholly volitional. However we can, if trained well, cope with and limit the effects created within ourselves.

The Thermoregulation System

The thermoregulation system that controls body temperature does it in two basic ways, firstly by the use of vasodilatation. This process involves the expansion of the blood vessels close to the skin, allowing for the raised heart rate to pump the blood quickly and towards the skin where it will cool, as physical exertion increases so does the heart rate. Secondly, perspiration is used, opening the pores of the skin, so that the body can sweat and cool down. When the body is operating in this manner, usually you will see the skin colour is red and oxygenated; blood vessels are wide open and the muscles full with blood.

Fear however is an evolutionary mechanism, designed to allow the human being to survive and reproduce. We can for a short time habitualise our response to fear, but we will never be able to make it extinct! That is not until the day comes when the whole of the human species live in perfect harmony, with absolutely no exposure to danger of any kind. So until that evolutionary day comes, our instincts will prevail.

When fear activates the human body the homeostasis balance is very different from that described above. The heart rate again increases to very high levels very quickly, but unlike the above, this time the rate seems to jump from resting to your heart trying to jump out of your chest almost immediately.

Vasodilatation occurs, but on this occasion, blood is taken away from the extremities of the body and distributed within the main organs and the major muscle groups. The body will sweat, but this time it will be cold and clammy while immune and digestive functions are inhibited (the fight or flight response). This all happens as the body prepares to deal with the perceived threat of danger or death. Not only do humans have the fight or flight response, we also have pre and post behavioural possibilities, these are posturing and submission.

Experiencing either fear or stress is a state of high arousal, one that can have a significant effect on performance "learning the psychological techniques to manage stress not only reduces discomfort but can enhance performance" Asken, (2010). An understanding of these states of arousal can help an individual to identify the cause and cope with the effects. Training in specific methods can reduce these effects and is something any serious martial artist should consider. There is a long list of stress, fear and anxiety symptoms, these are categorised into three basic representations: physiological (internal), physical (external) and psychological (mind) some of which have been covered above. These disruptions provide clear evidence that the body is subject to a cause and effect relationship with certain stimuli, resulting in either a general degradation or enhancement in performance levels. There is therefore a very real possibility that long term exposure to stress could cause serious injury to an individual, especially if their occupation involves repeated exposure to stress with no correcting strategies.

Goldstien & Koplin (2007). Consider the long-term physical or mental consequences of stress and the causal effects on the long-term allostatic load. Putting forward the idea that prolonged, intensive activation of effector systems could exaggerate effects of intrinsic defects in any of them; just as increased air pressure in a tyre could expand and eventually "blow out" a weakened area. It is not difficult to imagine that repeated or long-term stress or distress could lead to a medical or psychiatric "blowout." As a result of this, it is vitally important that individuals that have repetitive exposure to stress are given the correct tools to cope with the long-term casual effects of stress.

The next question should be, how could we counteract the maladaptive responses to enable an enhancement of performance? This will require mental force and the application of attention.

Experience is nothing more than a continual constant exposure to incoming stimuli in varied circumstances. This leads to knowledge of what has happened in the past and what may happen in the future. Coping with a state of arousal is no different; the more that you are exposed to a stimulus the more habituated you become. Experiencing a stressful situation will expose you to episodic memories, memories that are encoded through experience which are never forgotten, even though sometimes they are not the most welcome of memories. Psychological techniques that are designed to level out both physical and psychological reactions will need to be trained, to provide a coping strategy enabling a person to deal with high emotional feeling of stress and fear brought on by incoming stimuli.

Managing High State Emotional Arousal

There are a great deal of techniques which help manage stress or fear, these range from the most common breathing techniques, to visual imagery, mental or word associations. Learning one of these methods will help any individual manage behavioural reactions when the body has been exposed to a high state of emotional arousal. For hundreds of years there have been meditative techniques used to calm the mind and control the body. Taking a breathing technique and renaming it to something more adapt to modern combative, allows for practitioners to accept it as part of the training required to increase and control performance while in a high state of arousal.

Hyperventilation is a term given to a type of breathing rhythm that occurs when the body is going into a state of shock over something that the mind perceives is a threat. Faster breathing results in carbon dioxide being expelled from the body. This creates an arterial concentration of carbon dioxide, which in turn raises blood pH levels and leads to alkalosis. As well as this state of breathing having a causal effect on the blood within the body, affecting the homeostasis balance, it also affects us in other ways.

Chest pains, headache, body spasms, numbness throughout the body and ultimately fainting can occur. Fainting protects the organism from continued exposure resulting in damage to the body; it's like a re-booting of your computer when an overload occurs. If while in this state, the person experiencing the shallow increased breathing can be settled and encouraged to concentrate mental force 'pay attention to' their breathing and relax, taking in more oxygen in slow deep breaths, often they will soon come down, controlling both physiological and psychological body reactions.

Being able to control body reactions to stress and fear relates directly to classical learning 'Pavlov's dog' and the un-conditioned response to incoming stimuli that I discussed in the chapter 'To Think What Has To Be Thought. To be able to alter conditioning that has already been coded into the mind is a key goal to enhancing performance.

As well as there being a causal effect on the body which slows it down and helps protect the organism when fear or stress start to occur, there are also natural responses that heighten the body's reflexes and senses, it's important that these are also understood.

The cause and effect happens when the body receives a stimulus that is sudden, unexpected and severe. The mind draws attention to a certain stimulus that stands out from all others, thus directing the mind's attention towards the cause. Increased auditory awareness, visual clarity, quickened perception and awareness and slowing down of time all occur. What this indicates is that there is a clear difference in a maladaptive response to an adaptive one.

In research conducted by Leavitt (1972, 1973) cited by Grossman (2004), he used heart rate as a direct indicator to performance, at 115 BPM he noted a deterioration in fine motor skills, 145 BPM deterioration in complex motor skills and at 175 BPM a catastrophic failure in cognitive processing capabilities.

Relying on this research would lead to the conclusion that once heart rates reach 175 BPM it's all but over! However, there is research that contradicts these findings. Through themed analysis of introspective interviews post confrontational incidences and heart rate monitors during incidents, it has been found that the expected shut down did not happen. Heart rates in excess of 175 BPM were recorded while in the midst of deadly force encounters. "stress induces a heart rate increase in the area of 145 BPM, there is a significant breakdown in performance. But this is not true for everyone. Apparently, if you have practiced the skills extensively, you can 'push the envelope' of condition Red, enabling extraordinary performance at accelerated heart rate levels" (cited by Grossman (2004) page 34).

Practice is the general term used by instructors to encourage students to practice in their own time. However if we really want to make a difference instructors have to empower their students with knowledge of why and how to practice. Why?, should be obvious, but let's not make that age-old mistake of assuming that everyone knows the 'why'! I will come back to this point again and again and make no apologies for doing so. There should be no difference in the training method, of a professional police officer, military soldier, the door supervisor, close protection personal and Mr. Joe Average who wants to learn self defence. When I say method, I mean the teaching structure and the way in which it's taught. Yes the professional will train with real weapons, yes they will train more often, but the point I am getting at here is that there should be no difference in the value placed on any one's life! All should benefit from the most up to date training methods that are available today and be trained in a way that produces the best results, to do anything less is to fall short of your responsibilities.

To provide an example of individuals coping with high heart rates while exposed to fear, high emotional arousal, will lend evidence to the opposing view of Leavitt "a bizarre set of perceptual distortions can occur in combat that alter the way the warrior views the world and perceives reality. It truly can be an altered state of consciousness, similar to what occurs in a drug-induced state or when sleeping. It is amazing that we never knew about this before. All we had to do was ask" Grossman (2004). What implication does this have? Is it possible to train in a way that will enable any individual to use this state of consciousness to their advantage?

The difference between Leavitt's findings and the research conducted by Grossman is a marked one, with the latter group being capable of performing while under the highest levels of stress. What is it that allows these individuals to cope and perform when others fall short? Put simply it's experience. Experience to the stimulus that creates these high levels of arousal and with experience comes training in the correct manner. Experience within this type of environment requires continued and repetitive exposure to the stimulus that triggers high levels of arousal. Do not however fall into the trap that a job automatically gives experience, take any job where exposure to high stress stimulus can occur, for example, police officer, prison officer or army soldier. It's hard not to imagine that these occupations will result in exposure to high stress, it is so often the case that individuals in these types of occupations fall victim to physical and cognitive shut down. It's therefore lack of experience, or more accurately, lack of exposure to direct experiences that involve aggressive or deadly force stimuli, that in turn provide allostasis changes, which create the ability for individuals to perform while under high levels of stress and fear. Unless exposure to these stimuli happen on a day to day basis, you will fall foul of high heart rates, disorientation, tunnel vision and all the other mechanisms that the body uses to control itself while in the state of high emotional arousal within the regular routine of your day.

The correct type, method of delivery and training is therefore very critical when it comes to transferring knowledge and skills to any student, it does not matter if you are teaching occupational professionals or the individuals that come to class to learn how to protect themselves, they all deserve the very best that you, as an instructor, can give them.

Understanding stress/fear and the effects that it can have on performance, should then give us the knowledge to train in a manner that encourages constant exposure to a stimulus that induces stress. This is not as easy as it seems, although may be easy for some that are more inclined to heightened experiences. However, experiencing real danger is hard to replicate in a normal training environment. The most reliable way to avoid degradation in performance is to create experience, in doing so we also have to ensure that the correct muscle memory is encoded. Muscle memory is nothing more than another word to explain spontaneity of movement, not having to think about your responses to certain events. This has to be engrained so that performance at high levels of stress-induced heart rates can be maintained.

Can the human body perform under stress? Imagine yourself captured by terrorists, tied to a chair in a room, a team of special forces explode into the room guns blazing, life in the balance, do you want them to spray the room and hope for the best or pick their targets? In doing this action do you think they are operating under high stress? What makes them capable of these actions? and why can't you have similar control?

Do you lose motor and mental control? Yes, if you are not trained and experienced! and therein is the key, "trained and experienced".

So how does this help the mainstream martial artist?

Mindset and understanding are two of the hardest elements to teach within the arts. There is a saying "you react in the street, the same way you train in the school" or something similar. Training realistically and in a manner that creates reactive spontaneity is a key to effective defence. Experience is something that needs to be obtained, experience of efficient repetition, experience of your own emotions and feelings, experience of dealing with internal as well as external mechanisms. Knowing what your body is doing and why is the first step to creating coping strategies, which will enable effective and controlled responses when performing under a high emotionally induced stressful situation.

The aim of this knowledge is to equip an individual with the skills that will enable them to continually perform while under the emotions of fear and stress. Before I end this chapter I want to give a brief overview of a strategy and technique that is necessary to allow transference to real life encounters. From my perspective it matters not if you are a professional or Mr Joe Average, you deserve the best. Asken (2010) in his book Warrior Mindset, identifies several physical and psychological strategies to enhance performance, I want to explore briefly one of these training methods.

Stress Inoculation Training (SIT) was first researched by Meichenbaum, D. (1976) cited by Asken (2010). His research in cognitive behavioural training helps provide a method to enable an individual to cope with stressful situations, through a program of managed successful exposure to the stimuli. His program usually encompasses three phases.

1 Conceptualisation

This first phase relates closely to the mindful attention explored in the chapter ' To Think What Has To Be Thought' creating mental imagery that is vivid and positive rather than negative in any way, helping the mind create a positive mental attitude. To get to this point requires an understanding of where you are and what risks you will encounter. Knowing what the stimuli are that first bring forward the emotions of fear or stress is very important. Having the analytical ability to analyze these thoughts and emotions and then create a work around is a very important key.

2 Skill Acquisition and Rehearsal

Skills to counter-act stress are centered more on the mental process involved rather than physical skills. Your ability to control your breathing, self guided talk, relaxation techniques, access encoded training all help to support phase one. These are the practical skills that are internal, they are just as important as the rehearsal of the physical skills. Indeed one may argue that they are more important if, at the time of critical application the mental skills are not available, choking can occur and no physical skill in the world will dig you out of the black hole that is indecision and physical shutdown.

3 Applications

Finally we have the application of both the physical and the psychological skills into either a training based scenario or a real time event. The latter creates that all important element ' experience' this is not to devalue training scenarios, just a simple truth, nothing will generate fear like the real thing.

It should have become very clear that performance under stress encompasses the whole of the material in this discourse, every part has its role to play. It will be the case that personal identity will shape the needs of the character, knowing and understanding the processes surrounding enhanced or degradation of performance will give keys as to what each individual may require to get the best out of their performance.

Maximizing performance involves both the physical and the sensory, it's not just about how fast you can run, or how quick you can apply an arm bar to an aggressive person, your sensory recognition of environmental stimuli is just as important and should never be overlooked. A German diver, Tom Sietas, achieved the current record for the longest breath ever held, he held his breath for twenty two minutes and twenty two seconds in June 2012. Although this is an amazing physical achievement the control over his sensory input would have also had to be huge to enable such a feat. Imagine the complete control and awareness that he had of his mind and body while in the zone.

10 THE BULLY, AGGRESSIVE AND IN CONFLICT

Bullying the Age-Old Problem That Will Not Go Away!

Why is it that there is always someone wanting to steal your money, take your lunch or just pick on you because they feel they can, of course there are always those that have experienced success in fighting and then want to prove how big they are by inflicting pain on the geek, or socially timid child? As a child, experiencing the torment, either mentally or physically inflicted on you by a bully can be a very frightening event, leading to that child becoming withdrawn and socially excluded. Everything will suffer, their school work, confidence, relationships with their parents and within themselves, with feelings of low self-esteem and a lack of confidence, all of this could ultimately lead to long-term transference of anxiety into adulthood.

There are ways to fight back, but first we must understand this age-old problem if we want to give these children a fighting chance. While exploring this subject I will also be touching on knowledge and understanding that are wholly relevant to the subject of this book, evolution, psychology, why we behave the way we do, as the adaptive humans that we are and how martial arts can help in this very touchy subject, hopefully this chapter will explain and provide a further understanding of the human mind.

In the UK schools as with just about every school in the world today, know that bullying exists within their schools; yet the age-old problem of bullying has still continued generation after generation! There are bullying strategies in most schools and in December 2011 the UK Government published new advice on bullying for head teachers, staff and governing bodies. The document, 'Preventing and Tackling Bullying', is intended to "help schools prevent and respond to bullying as part of their overall behaviour policy." Updated in July 2013, it also outlines the Government's approach to bullying, the legal obligations and the powers schools have to tackle bullying and the principles which underpin the most effective anti-bullying strategies in schools. Also in the UK the new Ofsted framework, which came into force in 2012, sets out that schools will be expected to demonstrate the impact of anti-bullying policies. Department of Education (2013)

So why over such a long period of time have we not been able to eradicate this behaviour? Bullying is a distinctive behaviour pattern, which involves individuals deliberately harming and humiliating others, creating an asymmetry balance between individuals or groups. This type of behaviour is intended to harm or inflict psychological advantage, it occurs repeatedly over time, resulting in an imbalance of power. This behaviour can also be termed "aggression, conflict or competition" all of the above are fallout from a bully's behaviour.

Here is an extract from Kurtz and Urpin (1999) The Encyclopedia of Violence, Peace and Conflict "When we try to define aggression we often do so in terms of the motivations and intentions of the actor, and sometimes the recipient as well. We may define aggression as behaviour that is intended to harm another, or behaviour that deprives others of their rights, or freedom, against their will. Such a definition implies that we know that the motivation of the actor was selfish or spiteful, that the actor intended harm to the victim, and that the victim would rather have avoided the consequences imposed.

Of course, we may find that after we have classified an act as 'aggression', the victim denies that he/her was forced to do anything against his/her will and/or, that the perpetrator denies any ill intent, insisting that he/her had no intention of actually harming the victim, that his/her behaviour was for the victim's own good, and that the victim was a willing participant (he/her had 'asked for it') or deserved the treatment received.

The very existence of such arguments implies that the definition is subjective and judgmental rather than an objective description of observable behaviour". This description can be re-classified as bullying and relates in almost every detail to the motivations of a bully. But that's not all that is occurring, children, adolescent teenagers, adults, they all in some way seek a certain hierarchy and in doing so they create status, this may not seem very obvious at the moment, however later in this chapter evidence will be provided that supports the fact that the creation of "status" may well be a key reason for the underlying behaviour of the bully.

There are views around that that put forward the fact that bullies are not born, instead they believe that they are molded through their early developmental stage in life, through bad parenting of aggression traits starting as early as the terrible twos! Is it simply the case that if there were no victims, then bullies couldn't exist? As we all know that they don't pick on just anyone, they single out children with a certain lack of confidence, who give off a sense of fear long before they ever come into the bullies crosshairs, and let us never forget the fact that once a bully always a bully right? Well that's not the case, if it were, what you are saying is that no individual can learn, change, adapt and that they are pre-programmed from birth with no chance of any volitional choice. Adult bullies have a very destructive character and will bring fear and intimidation to groups, whole societies as well as the isolated wife manipulated controlled and abused behind closed doors.

Why Children Become A Bullies?

The first task here is to explore when and why children begin to develop the bully trait or behaviour, and is this behaviour culturally and socially specific to regions or countries? If it is not, and bullying is a universal global problem set across all types of countries and societies, then we have to entertain the fact that this could be a hard-wired human mechanism, left over from our caveman past.

To answer the second question above we can look at research conducted "this international cross-sectional survey included 123,227 students 11, 13 and 15 years of age from a nationally representative sample of schools in 28 countries in Europe and North America in 1997–98. The main outcome measures were physical and psychological
symptoms. Results: The proportion of students being bullied varied enormously across countries. The lowest prevalence was observed among girls in Sweden (6.3%, 95% CI: 5.2–7.4), the highest among boys in Lithuania (41.4%, 95% CI 39.4–43.5). Due. (2005). What this research tells us, is that bullying is not contained to any particular culture, society or country.

Every country surveyed contained instances of bullying from European countries, to America and Japan, providing solid evidence that bullying is ubiquitous across human cultures. It does not take long to find instances of bullying in every country around the world, so does this all stem from aggression problems in children that start around two years old? and if so why? What is clear is that bullies are born and they are not made. Every single person on the planet may well have the capacity to become a bully or to a lesser degree may already be one, if they are not a bully then they are submissive to all those that are.

There could well be reasons for this as well, it may be the case that humans have evolved with these two dichotic traits, a dominant or a submissive trait and this is the natural way of things. Not a thought that some may wish to entertain, as if this is indeed the case, then humans have a capacity for this very distinctive behaviour, it is genetic, inheritable and passed down through the generations and not as some may argue molded from the parental nurture or rather the lack of it. Instead it seems that this behaviour has its origins in stone age man, one thing is for sure we have no way of telling what behaviour was like at that time, did pre historic cavemen's children bully the more vulnerable among them to gain power and in achieving this give them access to more resources, attracting a mate in later life?

No amount of research into our fossil history will provide an answer, but could we find an answer in our closest relatives? If primates and other non human related animals have the same type of behavioral patterns this would help answer the question and provide humans with a far greater understanding of the nature of bullying. The problem here is the evidence used to show intent, how would we know that a primate has made a volitional decision to inflict harm to another primate in order to gain power and status over them? If primates have these same behaviour tendencies then this would effectively pre date the dawn of homo sapiens?

Primate Evidence.

It may be obvious, but primates do not use complex language skills like humans to communicate with each other, and as a result they lack the ability to create an understanding of 'reasons why', why did you hit me? what did I do wrong? I am receiving this present because I did a good thing! they use verbal communication at a basic level. Research has been conducted into primate behaviour, with interesting results "Bullying-like behaviours are found in every major group of primates, and can sometimes be severe. Among baboons, one of the best-known, non-human primates in the world, bullying-like behaviours are common" Altmann, 1980). Cited by Hogan & Sherrow (2011). Later in the same article they go on to say "Chimpanzees live in communities with many males and females and males live in the groups their born into their entire lives. Males also form dominance relationships with each other based on physical power and friendships, which they use competition over mates". Here we see the use of the word bulling, which as I remarked above could just as easily be aggression.

Aggressive bullying behaviour in Chimpanzees usually takes the form of non violent attacks, the most common form of aggressive displays involve the male alpha or challenger puffing themselves up, with all his hair standing on end and walking or running in the direction of the target, there have been recorded events where the challenger picks up an item that is lying around and uses it as a threat. The typical reaction of the subordinate male who is lower in status, is to use an expression of 'fear, lower his head, and put out their hand, palm up in a giving surrender posture.

Along with these body postures and actions the male will also grunt several times. This is a typical submissive behaviour and will ensure that no physical attack where injury may result, will take place. If the males are very close in their social status then the outcome could involve more intense interactions, they may lead to a very dangerous encounter. Chimpanzees have been known to gang up on former alpha males and kill them, an extreme violent behaviour indeed and the observed reasons for this behaviour was non-adherence to social norms within their own group.

Here is an account cited by Hogan, M. and Sherrow. PhD. (2011). On October 29, 2002 David Watts was observing males from the Ngogo chimpanzee community in Kibale National Park in western Uganda, when he observed a gang of adult males attack and kill a young adult male named Grapelli, from their own community. I had spent a lot of time with Grapelli over the previous two years, and had gotten to know him fairly well during that time. He was a striking example of a young male chimpanzee, with distinctive diagonal black markings on a rare, light tan face. He was also one of the biggest, most aggressive chimpanzees at Ngogo and didn't spend much time with the older, higher ranking males of the community. Instead, Grapelli would go off by himself, for weeks on end, and when he returned he would fight with the other males.

Between when Professor Watts left the party of chimpanzees on the night of the 28th and when he caught back up with them on the morning of the 29th, something had snapped in the other males. When he arrived on the scene, the attack was already underway, and a large group of adult males were repeatedly attacking Grapelli, pulling, punching, kicking, dragging and biting him, until he was bloodied and struggling for breath. Grapelli was beaten so badly during the attack that he could barely manage to pull himself into a rudely constructed nest in a low treetop before collapsing. The next day he was missing and it took another eight months before his decomposed body was discovered by two of the Ngogo field assistants.

What is very evident here is that behaviour of bullying within humans is designed to achieve the same results as is evidenced within primates and many different species of animals not so directly related to our ancestral history. Children, adolescents and adults in every country around the globe display traits that are now labeled as bullying. Behaviour that is a result of natural selection and survival of the fittest, has been turned up a notch or two and through the medium of language and cultural norms manifested itself as heightened un-natural behaviour, which in turn has been categorised as a social no-no.

Krutz & Urpin (1999) in their chapter on Animal behaviour studies in primates, talk about aggression modifying behaviour in different ways, affecting the aggressor and the recipient. A primate that is subjected to aggressive behaviour will in some manner have to modify their behaviour, this modification can take one of two paths, the recipient could stand up to the aggressor and in doing so risk injury, or they could submit giving up a resource such as food or a mate. This provides us with strong evidence that bullying or to re-phrase the word "aggressive behaviour" is designed to achieve dominance, is this behaviour universal throughout both human and primate species, making aggressive behaviour a trait that is hard-wired into our genetic makeup? Aggression in itself is somewhat subjective and depends upon the context that it is experienced, it does not require physical injury to be defined as aggressive, indeed the act of aggression can have a different outcome to both perpetrator and recipient than we first perceive, the subject of aggression is covered in the chapter on behaviour. Aggressive behaviour is designed to ensure that dominance over social groups and individuals is achieved, and in doing so, the dominant individuals gain popularity, they have more status, access to resources enabling them to survive longer and reproduce, in Darwin's term it's survival of the "fittest", therefore transferring their genes to future generation, ensuring their genes survive. Evolution has a large part to play in the majority of human behaviours; it's finding that evidence that leads to a greater understanding of why we humans, sometimes do what we do.

Other species also display such behaviour, take a litter of dogs for example, out of a large litter there is always one runt of the family, the small one that is always last to get it's chance to suckle mum's milk, if the runt is fortunate to survive this early stage it can then be observed being group bullied into leaving the food alone until the pack has had it's not so fair share and there is usually a dominant one or two that lead this behaviour. This behaviour is no different to humans or primates; there is a hard-wired mechanism that says, get more food and you will benefit above the others, even if they are members of your own family.

Cultural Reasons for Aggressive Bullying

Humans today are fundamentally different from our past ancestors, we live in a very different world, one in which global communication happens in a heart beat, we celebrate an event, post it on Facebook or twitter and a moment later someone reads that in a different country the other side of the planet, what would have taken a great deal of effort is just the click of a button away. We communicate via language that, to our caveman relatives, would probably sound like an alien dialect. Now consider for a moment that this internal mechanism that creates our behaviour, designed initially to allow for dominance and more survival opportunities is suddenly transported to 2013, the need to fight and use behaviour such as aggression to obtain resources to survive has now no real destination or outlet at least not in the primitive sense, it now manifests itself in today's young children as bullying behaviour, what we were initially hard-wired for has nowhere to disperse.

Communication has heightened our ability to transfer aggression, to speak words that cut like sun rays through the sky, humiliating and harming the subjects of our discourse and all this from the mouths of our most treasured possessions, our children. Not only do we have the medium of language, we also have a plethora of other channels in which to convey our hatred to our intended target and all with a certain amount of anonymity. The use of text messages, social media sites on the web, all allow easy access to intimidate, spread gossip and lies about our victim, while at the same time maintaining distance, it's no wonder that the behaviour once designed in our evolutionary history has now been distorted beyond all recognition, allowing for this behaviour not only to maintain itself, but also to change and become more intensified than it ever was in our past history, this being the case, the age-old problem seems destined to be with us for some time to come.

Culture has also helped mould behaviour that further supports bullying, in today's environment violence is seen almost every day, it may be the local news broadcasting images of police dealing with local criminals, television series portraying family violence, the local newspaper, internet headings, sports events like boxing, games that encourage children interaction with high-end violence, the list goes on, all of this helps support the view that what is termed social violence is acceptable behaviour "the murder rate in America today is six per 100,000 per year. If six more out of 100,000 people were convinced to kill, the murder rate would double, remember, murder is just the tip of the iceberg because for every homicide there are tens of thousands of injuries assaults, hundreds of thousands of thefts, millions of acts of bullying, and an untold amount who live their life in fear. The June 10, 1992, issue of The Journal of the American Medical Association (JAMA), the world's most prestigious medical journal, reported that violence depicted on television "caused" (caused is a powerful scientific word) a subsequent doubling of the homicide rate in the United States 15 years later. The AMA is so convinced of the impact of violent media, that they said if television technology had never been developed in the United States (or if we had kept our kids away from it) there would today be 10,000 fewer homicides each year, 70,000 fewer rapes and 700,000 fewer injurious assaults" Grossman (2004).

This is damning evidence that media based violence has had a significant effect on our children, changing the way that they view the consequences of such acts and in doing so developing a behaviour pattern that accepts social violence as a society norm! The box in the corner of the room has above all other media, been responsible for the dissemination of violence to the human species, what once started as an unassuming black and white TV screen back in the late 1920 has over the years transformed itself into a flat screen 3D high definition colour experience that provides the most vivid of stimulus inputs into the brain. Now add to this the growing popularity in reality TV programs which portray aggression and violence as an everyday experience and we have a caveman child being exposed to behaviour seen as social and normal, program makers see the inclusion of violence as an essential ingredient for entertainment.

When will parents realise this simple fact, it's not enough that all too often children have to be witness to extreme levels of violence within the family unit, they are then subjected to a near constant stream of aggression and violence, which 1,000 years ago would give a natural fitness advantage to those children that progressed through adolescence to adulthood, giving them a far better chance of survival, procuring a mate and distributing their gene for inheritance into their children. In today's society there is no requirement for this aggressive behavior.

It must also be remembered that bullying tactics are not just the sole preserve of the individual, groups, organizations and whole countries can become susceptible to the power struggle, resorting in all types of underhanded and big brother bullying tactics. These groups will target others with tactics designed to intimidate, coerce or harm them, all designed to ensure that the group achieve what a minority perceive to be, what the majority want! Political groups and leaders of countries use bullying tactics to maintain social order, to go to war, to suppress uprising, even though the mainstay of the population may disagree with the pattern of behaviour, they are usually powerless in the face of such power.

This seems a good place to look at something that up to now has only been hinted at and that's the creation of social status. One of the primary interactions of children from a very early age is that of negotiating a hierarchy and creating a status within their social group. It could be argued that the reason for dominant adaptive behaviour "bullying" is solely designed to elevate individuals to positions of dominance to create status within the group.

The status of an individual can be linked to a better chance of survival, more food, and prospects, all ultimately leading to survival of the fittest. However, a question should be asked at this point is, is hierarchical and status developing behaviour universal? Does this same behaviour such as identified in the primates, pervade other species as well? If it does, then just like the primate research above, it could answer a great many questions.

Researchers have turned their attention to social hierarchy, with a view to understanding this behaviour, and why it exist; Is it found in all human groups, including ancient tribes and ancestors, or across species? In his book Moral Animal, Wright (1994) talks about social hierarchy in hunter gatherer people, the Archay of South America are a traditional hunter-gatherer tribe living in eastern Paraguay, the best hunters of this nomadic people, provide meat for the whole tribe and in doing so provide valuable resources to those not capable of doing so, this seemingly generous behaviour has an underlying benefit, which is not at first obvious, the best hunters have more affairs and more illegitimate children, than lesser hunters, consequently their children get treated with more reverence than other children, this creates a higher status within the tribe for those best able to hunt food.

This hierarchical behaviour becomes even more evident when we explore the dominant world of a chicken, Shimmura, T. Eguchi, Y. Uetake, K. and Tanaka, T. (2007) explored the social relationships between members of a poultry group. There were two basic categorised social actors, one was the subordinate and the other the dominant, thus creating a dominant hierarchy. The subordinate does not engage in any aggressive behaviour, instead they do their best to avoid any confrontation with a dominant.

When food was restricted and then introduced aggression became commonplace. The aggressive behaviour identified was, aggressive pecking, displacing, chasing and threatening. Aggressive pecking was noted, and recorded as either being severe, forceful or light, this enabled the researchers to identify which hens were the dominant birds and which were the lowest subordinate. Creating resources, such as an enclosed nest site, a raised perch and a dust bath, led to increased competition between them for these resources.

This research demonstrates the hierarchical dominant status and pecking order of the hens, which again lends weight to evidence that status in social groups is an adaptive mechanism allowing for the survival of the fittest, which is universal across animals. What is important here is that there is no conscious thought on behalf of any chicken that they have to engage in aggressive behaviour to move up the pecking order. The subordinate chickens already have a behaviour which dictates that they are not going to beat the dominant chicken to food and fighting will only cost to much recourses, better to wait my turn. It's not the pea sized brain that is doing the thinking, its already been predetermined by the genes. The genes have, through many generations adapted behaviour to favour survival.

How does the behaviour for status link in with aggressive behaviour of children, when they bully another child? Status is not only gained through aggression, strategies and skill play a role as well, a clear thought process to help explain this behaviour was put forward by Wright, R. (1994) Moral Animal. "for decades while many anthropologists have down played social hierarchy, psychologist and sociologist have studied its dynamics, watching the facility with which members of our species sort themselves out. Put a group of children together and before long they fall into distinct grades, the ones at the top are best liked, most frequently imitated and when they try and wield influence, best obeyed, the rudiments of this tendency are seen among children only a year old. At first status equals toughness, high ranking children are the ones that don't back down and indeed for males toughness matters well through adolescence, but as early as kinder garden some children ascend the hierarchy via skill and cooperation" this toughness and wielding of influence are the same behaviour traits that we associate with the bully or the child that stands up to the bully, by default they also move up the social ladder in status, creating a reputation for themselves as also being tough.

If this is indeed the case, are there any true egalitarian societies or have there ever been any? Egalitarianism is a theory that there are societies or groups of animals that have an equal amount of equality/status. Supporters of this theory state that all humans are equal in fundamental worth or social status, this would have to be questioned in light of the pecking order created by hens, let alone the behaviour of primates and humans. It makes sense that a political organization may want to promote a egalitarian doctrine that everyone should be treated as equals and have the same rights, or the same social standing, or economical value, as their goal may make for that particular party being re-elected. Some even define egalitarianism as the point of view that equality reflects the natural state of humanity, so let's not confuse politics for an evolutionary adaptive mechanism designed to aid natural selection.

I spoke above about the inability for some parents to cope with early aggression in children and that this could start as early as two years old, the behaviour of children at this early age, seems to be hard-wired into the child from birth, as most parents would attest to experiencing this type of behaviour in their two or three-year old children. Up to the age of two a child has usually had no real input as to aggressive behaviour patterns, is this the age when these evolutionary traits begin to reveal themselves? Or is it the parent's fault for mis-treating the child at home, causing feelings of insecurity, which leads to aggression? Observed behaviours that children witness may well play a role in supporting the inbuilt behaviour patterns that society now refers to as bullying.

The Terrible Twos

Every parent gets to hear about the terrible twos and how behaviour during this time can be a testing one for any parent, it's the beginning of a new era when the unassuming smiling, crying, loving baby begins to show patterns of behaviour that are testing to a great many parents, some say they never experience this, but those that have are often unprepared for the change from a cooperative baby, to nothing is right! This change is not timed exactly with the age of two it can start before two, or later, up to the age of three. During this stage cooperative behaviour changes, as a parent you may notice a new level of NO! Doing the opposite of what you have told them "don't throw your cup on the floor" and guess what it ends up on the floor so many times you eventually take it away. Tantrums are common place and can occur out of the blue and for apparently no reason, at least to the parent, you suddenly have on your hands a more assertive bundle of joy.

At around this age, the brain is going through a massive amount of development, the mind is beginning to have an influence on what the child perceives to be what they want and not what you want for them, they think for themselves and begin to demand attention, they want things that are important to them and no amount of logic from the parent can persuade them otherwise. Feelings, language and increased mobility all start to occur around this time, bringing with it the ability to communicate, express emotions and move independently. The stage prior to communication, via the medium of words is termed pre-symbolic; when the child has the ability to speak it's called the symbolic stage. During this time they have to deal with a fantastic amount of in-coming stimuli, their brains are undergoing neurobiological and psychological changes, resulting in self-awareness and a sense of understanding of self.

What impact on behaviour in later life does this stage of development have and can this early transformation, if handled badly by the parent, result in a natural tendency for bullying? One of the traits displayed is that of an unwillingness to share, they are possessive of food or toys, this in itself could be a very early indicator that what they have they need to retain, in order for them to nourish themselves and have tools to explore. Learning to interact with other children also becomes a key skill throughout this time, as up to now the majority of contact has happened within the immediate family, they now begin to mix and interact with other children, this time of early interaction is crucial to learning acceptable social skills "researchers have found emotional development and social skills are essential for school readiness. Examples of such abilities include paying attention to adult figures, transitioning easily from one activity to the next, and cooperating with other kids" Cherry, (2013).

These points are key if we are to find any link between this stage of developing behaviour and patterns of behaviour in later childhood and the point that is of great interest is "paying attention". In the chapter "To Think what has to be Thought", I explore in great depth the importance of volitional attention and how this impacts on our ability to wire the neurological pathways within the brain and how paying attention is a key aspect in the way the mind learns. For an early child, paying attention has the benefit of clear and unrestricted access to the brain.

Imagine an empty piece of paper just before you begin to draw, you are free to create whatever image you want! and that is what paying attention to parental behaviour does for the young child, they begin to recognise how to express feelings and interact socially, all through observation of the people they see the most, their parents. These observations imprint neurological pathways within the young developing brain and in doing so create patterns of behaviour.

It is not hard at this point to understand that if your child sees two parents at odds with each other, not sharing, or not correcting behaviour when the child does not share, that this will develop within the child an understanding that this is acceptable behaviour. If the innate hard wiring suggests that food needs to be kept, parents sharing food help suppress a heightening of this behaviour.

Parent interaction between themselves will also be attended to by the child and parents who argue, create tensions, do not show gratitude, or worse still shout, argue or fight will imprint this onto the child, who will then continue with this behaviour in a great many cases when they begin their interactions with other children.

If there is an underlying tendency for a child to dominate due to evolutionary adaptive mechanisms, they will learn that aggressive behaviour is the way to achieve this, they will also use subterfuge, lie, manipulate, influence and a host of other means to achieve their goal. Therefore modeling appropriate behaviours that are socially acceptable are essential at this stage. A well-known method of learning is operant conditioning; this type of learning involves either positive or negative reinforcement. This is where behaviour is rewarded, this type of learning also has to be considered carefully as reinforcing a maladaptive behaviour could be disastrous, that is why it is so important that when a child demonstrates positive behaviour that they are emotionally rewarded, which in turn reinforces the behaviour.

Cherry (2013), goes on to say "parents can also boost empathy and build emotional intelligence by encouraging their children to think about how other people feel. Start by inquiring about your child's own feelings, asking questions about events in your child's life. "How did you feel when you lost your toy?" "How did that story make you feel?" Once children become skilled at expressing their own emotional reactions, begin asking questions about how other people may feel. "How do you think Nadia felt when you took away the toy she was playing with?" By posing such questions, children can begin to think about how their own actions might impact the emotions of those around them". Emotional intelligence is a key factor in ensuring that any child begins to understand feelings, this also includes the natural feelings of fear and aggression, as it is these feelings that lead a child ultimately down the path of a bully, this is not to say that feelings of fear and aggression are bad, they are not, its managing these feeling and coping with them, while interacting within a social group that is important.

As we discovered above, evolution plays an important hand in providing adaptive behavior that in our caveman past was absolutely necessary, there were no child play groups available to these ancestral children, where they could learn in a controlled environment. With this in mind it's no wonder that today's children sometimes have a hard job finding the boundaries of what is acceptable behaviour. Now that we have began to understand the behaviour that we label "bullying" it's time to explore what we as parents or martial artist can do to help a child deal with this pattern of behaviour.

Born and Not Made

It has become evident that a bully is born and that they are also molded through two very powerful influences, evolutionary adaptive behaviour and parental guidance. Learning how to fight back is where I now turn my attention to, however it's really important that we learn from the evidence above that bullying is another word for aggressive natural behavior designed to achieve dominance, status and fitness over others, we will never eradicate it from our schools or our social lives as adults, what we must plan for, is education, helping both children and adults manage this type of behaviour and understanding our evolutional history.

It does not take a great leap of imagination or foresight to understand that children mimic behaviour that they see in both their peers and their parents, it is therefore vitally important that parents modeling aggressive behaviour, understand that their children are in all likelihood going to learn that very same behaviour. Even the words that they use to convey information about how to deal with aggression and handle physical contact are going to impact heavily onto the child's mind.

According to the National Society for the Prevention of Cruelty to Children NSPCC (2013) 31,599 children called their helpline in 2011/12 regarding bullying, 18% said that they would not talk to their parents about being bullied and bullying was the main reason why all boys called the helpline, they go on to say that 46% of children have been bullied at sometime during their life and that does not include the ones that would not admit to being bullied.

In researching the statistics on bullying one such number jumped out at me, 60% of those characterized as bullies in grades 6-9 will have at least one criminal conviction by age 24, according to Utterly Global. They state that the resource for these figures were, Bureau of Justice Statistics-

School Crime and Safety, Centre for Disease Control and Prevention and U.S. Departments of Education Associated Violent Deaths in the United States (1994-1999).

However there is no specific research that identifies this high figure, is it the case that behaviour designed to enhance status and dominance later in early adulthood transforms itself into criminal behaviour? There is no large body of research on this transformation, it may well be the case that this high percentage is a limited survey and designed to install fear into school authorities to make them sit up and listen, certainly more research needs to be done regarding the long term transference of early bullying behaviour into criminal acts.

At such young ages any behaviour that a child exhibits is generally for a reason, we may however sometimes struggle to ascertain the reason why at the time. Young children will display mimicked behaviour well before they begin to mutter their first words, parents lead the way by touching their head and encouraging the child to do the same, attempting to get the child to understand 'head'. Then ever so gradually they begin to develop curiosity and start to explore the World according to them. As the child grows older, more complex behaviour starts to reveal itself, copying parental behaviour by using a knife and fork, trying to put on clothes or performing body and facial expressions, all this requires is an amount of attention and they are off.

Play also develops with other children or substitute children in the form of dolls or that stuffed teddy bear, these are very early signs that children are beginning to imitate parental behavioural patterns. Kurtz, L, R. and Urpin, J, E. (1999) in their book The Encyclopedia of Violence Peace and Conflict, state that "there is no reason to believe that the process of learning differs among youth who are violent and those who are not" they support the view that learning happens through observation within the environment around that particular individual. Learning is reinforced by mechanisms of rewards and punishments. By far the most important situation and context of this learning process is close parental relationships, family and peer groups. However humans do not just play the role of mimic within this learning process, they take an active role in adapting their own behaviour.

They further state that "this can be achieved by the capacity to use symbols - fundamental to the capability of forethought - as well as self-regulatory and self-reflective capabilities. Self-regulation is more difficult for some children than for others however. For children who are impulsive, and for gang members and homeless street youth, for example, self-reflection and self-regulation are especially difficult, either because they lack sufficient internal controls over their own behavior or because of peer pressures and group processes".

What this provides us with is more evidence that early childhood behaviour progresses through to the adolescent youth and depending upon their cultural place in society could result in them being exposed to continued violent encounters. The opportunities for observing and learning violence are more prevalent within the environment of the street, especially areas that are saturated with gangs, homeless youths or poverty-exposed children.

Fast forwarding a few years we now find a child well versed in a set pattern of behaviour that can be traced back to early experiences. They are now entering the social world of interaction and have a minefield of emotional and physical interactions to negotiate. Let's look at a few thoughts that help explain why a bully bullies and why they continue to bully throughout life.

Once a bully inflicts pain and humiliation on their victim, they realise that they have power over others and like a drug, they feel good on it. This power also brings with it social status.

Nobody actually deals with them and tells them that it's wrong to bully, so they continue to inflict pain, as the behaviour is left unchecked they think it's ok to continue.

Bullies bully because they have low self-esteem, they feel insecure and are not like normal kids, as they do not have many friends and feel bad about themselves.

Bullies are psychologically damaged, either at birth or have become that way due to bad parental guidance.

Bullies have been made by their parents and have been exposed to violence and aggression within the family.

Not all of the above statements are an accurate representation of the facts "Research indicates, for example, that toughness and aggressiveness are important status considerations for boys, while appearance is a central determinant of social status among girls (Eder, 1995 cited by Espelage and Holt (2001). They then go on to say" Therefore, it is likely that this pressure to obtain peer acceptance and status might be associated with an increase in teasing and bullying to demonstrate superiority over other students, for boys and girls either through name-calling or ridiculing".

Research indicates that bullying behaviour is not about the bully fulfilling a need to harm and make afraid and in doing so satisfying a deep need for evil, although this may be the case in the odd child, instead it points to social pressures and peer group standing as one of the main causes for this behaviour "the analyses in the present study of 6th through 8th grade students quite clearly indicate that students who bully their peers on a regular basis share the same amount of popularity or peer acceptance (i.e., number of friends) as those students who do not bully their peers. This finding suggests that students who bully others are not necessarily socially rejected but do have friends" Espelage and Holt (2001).

This would also lend evidence to contradict the claim that the bully is insecure, has low self-esteem or has been psychologically damaged by its parents. Instead the opposite is maybe true, they are intelligent, strong and have a clear identity and sense of self, they are also supported and encouraged by their peers, they mix with children of similar traits, even though they may have been taught that this type of behaviour is wrong, these are very powerful supporting groups that will continue to encourage this behaviour. It's not all about social states and peer groups, bullies can lack self-confidence, or they desire attention and these feeling will have been exaggerated through the lack of early parental guidance.

Not one of the above reasons for why a bully bullies says anything about reciprocal altruism designed to create status hierarchy and how these traits support natural selection. Wright, R. (1994) Moral Animal. Puts forward the view that there are times when it makes good evolutionary sense to have a low opinion of yourself and to share that opinion with others, the whole reason for hierarchical status is that some genetic heritance will be able to transmit a mechanism that tells an individual, whether that is a human, primate or a chicken, that there are some among the group that to challenge would exult to high a cost and that it would be better to subjugate and preserve ones current status.

Wright further states that the method used to deliver this innate idea that the organism has a chance at winning, or thinks that there in no chance of success, is transferred to the conscious mind via feeling, you either feel confident about your success or you don't. Those that are at the bottom of the pecking order get these feelings more intensely, these are the feelings that can turn chronic and begin the long road to a feeling that could be called low self esteem, lacking in confidence, shyness or an individual labeled as quiet. This could indeed be evolutions way of allowing an individual to come to terms with the fact that they are of lower status to another and that they should be submissive rather than a threat. They are therefore perceived by a more dominant individual, as not being a viable threat to their high status position and are therefore not worth an amount of effort to engage with.

Its in their genetic interest to be submissive, they display this trait overtly so as not to be mistaken as a threat, those that do display a degree of self confidence may well attract the attention of a dominant individual, they may not be ready to challenge and therefore display submissive behaviour to mislead the alpha until such time as they have climbed closer to the hierarchy and status of the leader, or bully!

What's Happening Within the Brain?

It's only recently that researches have started to scan the brains of both those being bullied and the bully, with surprising results. When interviewed children who were on the sharp end of a bully's tactics reported the same feelings and symptoms that were given by people suffering from depression, anxiety and fear. This would suggest that there would be a manifestation of psychological and physical effects on the child. Researchers are now becoming aware of the true implications of bullying and how it affects children and their brains "using SPECT brain imaging, Todd Clements, M.D., Medical Director at the Clements Clinic in Plano, Texas, and I have discovered that the brain scans of bullied patients resemble the brain scans of patients with Post Traumatic Stress Disorder (PTSD). Patients with PTSD report identical symptoms such as depression, anxiety, insomnia, inattention, flashbacks, etc.

What this means is that the human brain is interpreting the trauma of bullying in the same ways a soldier's brain interprets the experiences of battle or how a car accident victim's brain interprets the accident. Humans have the ability to adapt to their environment, which gives them the best chance to survive. Unfortunately, a bullied child's brain interprets the bullying as a threat and adapts to deal with the trauma" Divine (2010).

It may well be the case that most of those that are bullied during childhood go on to be confident individuals in adulthood, however there are some that are severely damaged by the actions of a bully. If anxiety is allowed to transfer itself to adulthood, then changing long-term ingrained behaviour will be even more difficult to alter. This behaviour is in all likelihood part of normal children's behaviour patterns; its effect is exaggerated due to social and cultural changes throughout our evolutionary development. Knowing this should give parents and those that are in positions of authority in teaching children an advantage, to develop programs that can help both bully and victim.

Brain scans on the bully also revealed interesting results "that bully beats you up because he enjoys it. Healthy kids' brains respond to other people's pain with sympathetic twinges in their own pain centers. But bullies who witness pain show activity in their brains' reward centres. Aggressive adolescents showed a specific and very strong activation of the amygdale and ventral striatum (an area that responds to feeling rewarded) when watching pain inflicted on others, which suggested that they enjoyed watching pain "Newitz, A. (2008).

The activation of reward centers does indicate that some bullies derive pleasure from the activity of bullying, however this is not the case with every person that bullies, or at least from some interviewers research, there are children that state that they do not get pleasure from it. Further research needs to be done to confirm this as, even though on one level a child may not think with their conscious mind that they enjoy it, something completely different may be occurring within the brain.

It is also the case that a large amount of children who get bullied never find the courage to say anything to either parents or teachers, as they themselves feel that admitting this to peers is a sign of weakness. It is therefore up to those that are in positions of authority with children to be mindful of the signs that a child is being bullied. It's also important to remember that this is abuse, it may not be as bad as sexual abuse, but it's no less harmful and so the responsibility falls firmly at the feet of parents, teachers (martial arts instructors), friends anyone that has a child's best interest at heart. The signs may be very subtle and these include; signs of emotional distress, nervousness, anxiety, being withdrawn, tearful, aggressive, depressed, nervous habits, lacking in confidence, bruises or scratching on a young person or attempts to hide physical injury, torn or damaged clothing, missing personal items, unusual bed-wetting, fear of going to school – (excuses of illness often made to avoid going to school), coming home without money or belongings that they should have, having trouble with schoolwork or grades for no apparent reason, lack of interest in doing things they would usually want to do. All of these signs should be taken notice of and monitored.

Behaviour clues are only as good as the person paying attention to the changes that are happening within the child, for a child to hide stress, fear, or anxiety and possibly physical injury, bruises or scratches for example will take a large amount of effort on their behalf and in the majority of cases they may not even recognise the change themselves. An emotionally insecure child, that suffers from the effects of a bully, which we now know range from anxiety to chronic depression will have their whole life affected. The inner strain and psychological effects will take their toll and lead to external manifestations in behaviour.

I introduced the concept of self-deprecation above, when I considered the evolutional reasons for this type of behaviour. Given what we now know it's no wonder that a child victim of a bully would begin to think in this way and may begin to display these internal thoughts in the language that they use, they may on the other hand develop coping strategies that hide their inner most thoughts and worries, that is why sometimes identifying the victim is harder than you may think. The mind has the ability to adapt remember.

A Plan of Action.

It's important that we all remember that fear, anxiety and stress create changes within the child, the brain adapts to the psychological threat, therefore we will need to focus on the mindset of the child, superficial patches that deal with outward behaviour will only create short-term results. To obtain long-term recovery the mind will have to be rewired back to a more stable setting. To help address these underlying deep issues, principles of cognitive behavioural therapy (CBT) should be part of your tool box, martial artists have been practicing this for centuries, it's know as meditation. Children will also need coping strategies and skills that will help them manage interactions with a bully.

Those that are in positions to help children need to have an idea of how to interact with the child, what they should do and say, there are some things that should not be said or done, especially if you want to avoid making a bad situation even worse for the victim.

1. Ensure that you remain calm at all cost, the child will see you showing signs of anger or frustration and focused on this, they may believe that they should not have said anything.

2. Keep reinforcing that they have done the right thing by bringing this to your attention, let them know, that you know it must have been a difficult decision.

3. Take time to discuss how they feel, slowly getting around to the important questions, who, where and when?

4. Take it slowly, one step at a time; the longer that the child has been bullied the more emotional torment has occurred.

5. Agree with the child the first steps, it should be their idea if possible and not yours, working together to tackle the issues that are raised.

6. Are there simple ways in which avoidance of a person or place could stop the bullying, keep in mind that once a bully has identified a target they may go out or their way to find them and continue the abuse.

Do not leave the situation without dealing with it, thinking that it will sort itself out; take action before things get any worse. Tackling this behaviour needs to be with the support of schools, clubs or any event that your child attends. Make sure that you meet with the person at the school who is responsible for their bullying procedures, they should have an anti bullying policy, make sure you see it. Get someone to take responsibility for monitoring interactions while at school. Remember recording meetings and agreed action in writing places a degree of importance on the matter, it's not going to go away and needs to be dealt with. If you are not happy with the way things are going keep going up the line of accountability, board of governors, positions, you are the only person that can keep the ball rolling and protect your child's long-term emotional intelligence. This all seems logical information and hopefully it's not new to parents.

Martial Arts

Any good martial arts school will have a policy in place to protect the children that attend, however it's not just the policy it's the whole mental and physical training that is important. Good martial arts' instructors will teach children non-violent solutions to bully avoidance, this will include strategies to talk and avoid a confrontation, and it's about confidence in yourself that you do not have to resort to physical confrontation to deal with the bully.

The goal is to discourage bullies if the potential victim can resist the verbal assault taking away the control and emotional pay-off, the bully will be less likely to choose them again, as with most perpetrators of violence or crime they are looking for easy individuals to attack. One that has the potential to fight back or is aware and not an easy victim will probably not be chosen. Tactics that are taught included avoidance, appropriate verbal exchanges, neutral/confident body language and facial expressions, selective ignoring and self-control, physical confrontation is the very last thing on the list. Bullies are looking for the nerve that makes any individual react, if a child can hide their emotions then they are well on the way to counteracting the bully. Once they discover your weakness they will do it again and again to see the same reaction over and over again.

Physical techniques are the last resort to dealing with a bully; the problem with any conflict especially with children is knowing were the boundaries are, when is it the right time to act, to fight back? Is there a time? Should you never resort to physical violence even when being beaten on? There are some that would support this, with words such as "if you use violence against a bully you are then no longer any better than the bully" we have all heard this at some point I am sure and to be fair adults have the same problem, when to fight or when not to fight? Should I use a pre-emptive strike? These are important questions and ones that any good martial arts instructor can help any child with.

The most important point is should any child use violence to protect themselves? The answer to this question will be different for a great many individuals, ask an adult this very simple question, if you were being beaten by another adult and had the means to defend yourself would you? This chapter has focused on the behaviour of children; we should by now understand what is happening on both sides of the coin, everyone has the right to defend themselves and their family, should it be different because they are a child?

Children will eventually come of an age where they can process this line and do the right thing at the right time, the main aim here is to convey knowledge of the bully and the victim, what they are and how to help a victim fight back, hopefully before too much psychological damage has been done. If you are dealing with a child who has already become a victim, then its no good just treating the behaviour of the child you have to get to the root cause of the problem the psychological effect on the mind, you have to know how the brain and the mind are working to be able to provide practical and effective techniques to help the child.

Be Mindful All The Time!

Cyber Bullying

During the first half of 2013, a young 14 year old girl took her own life due to Cyber bullying, if this tells us anything it should be, that parents have to be vigilant all the time. This type of bullying is disturbing and it's not the first time that young adolescent children have resorted to suicide. Cyber bullying is perhaps the most dangerous type of bullying directed at children, as it can be done anonymously and has a very powerful effect on the mind. According to the NSPCC website (2013) they cite government statistics on bullying, collated from government reports and research, they say that 38% of young people suffer from cyber bullying, this identifies a significant media for the continuation of this behaviour. Why would someone believe what he or she read on an Internet forum or text message sent to his or her phone?

The reason is how our brains work with regard to the bias law of similarity. If it looks like a tiger, it is a tiger. Like causes like and the brain perceives this, if we see a person being sick just after eating a particular food, we ourselves will not want to eat the same food, this has an evolutionary benefit, as it would have protected the individual from consuming the same food and suffering the same fate. Experiments have shown that if we create a negative thought and feeling then this will transfer to our conscious mind and become prominent, over the fact that we know it not to be true. For example take a glass of clear drinking water, apply a label to it that says 'contaminated with radiation' and feel the effects that this will have on thinking about drinking the water.

This rule of like causes like, can be seen when we observe an individual that has, in their past, decided to ink their body with 'tattoos' for example, or someone who has worked out and is big and muscular, we link these individuals with bad behaviour or crime and in doing so our minds automatically create a thought of, stay away from them, don't talk to them, there is a threat there somewhere, even though in our mind, if we take the time to think about it, there is no real danger. These linked thoughts of similarity are ancient wiring processes within the brain that are automatically transferred to the mind and brought into conscious thought. In the time of our Stone Age ancestors this process would have worked perfectly, there is a tiger, looking for food; we had better be on our way before we are the food!

Children these days are exposed to constant stimuli input from the cyber world, over time this informational input which was once used to aid humans' survival, has adapted itself to create the idea that what is read and spoken about on the internet is indeed "true and real", the mind uses similarity and adapts the belief within the individual that what they read is how it is. Cyber bullying is found in mediums such as email, text messaging, and social networks such as Facebook, My Space and "ASK", the last one being particularly unregulated. Here the bully can remain anonymous as they can create false names and profiles. Cyber bullying consists of the same threats that can be found in any bullying situation, with threats of violence, verbal abuse and the use of language that may not be normally said when face to face.

Anti-Abduction Skills

One of the most terrifying thoughts for a parent is the abduction of children, although a little off topic it has to at least be considered, although the statistics of child abduction is quite low, it's still a real fear. There are some techniques that children should learn in the event of a kidnapping attempt. Before we begin to think about dealing with this, we have to accept that this may happen and then
become aware of the potential danger, sticking your head in the ground like the proverbial Ostrich is not going to help. The first and most important tool is the use of the voice; this does require a great deal of confidence and is not the easiest thing for a self-conscious child to learn. Followed by distracting techniques to enable escape and lastly physical techniques that again aid and give time to run. This all sounds very logical, but a word of caution here, these techniques are hard enough for an adult to perform let alone a young child or adolescent teenager. A martial arts instructor teaching this requires a solid understanding of the methods needed to teach these techniques.

The Bully Who Knows No Right From Wrong

There are a few bullies where no amount of therapy will help, the only explanation is that the bully is a cruel individual, they like to harm and inflict pain on others. What's more, these bullies have no understanding of what is right or wrong, they feel no remorse, lack empathy and in most cases these individuals carry their behaviour with them through into adulthood.

This type of individual has a predisposition to violence, aggression, manipulation and lying, they can also be very intelligent, charming and in some cases very hard to spot, one in 25 adults have this type of character trait. Quite often people will use the term nature versus nurture, in the case of these individuals it is nature that has created the shortfall in ability to understand and nurture will only have a very limited effect. Understanding this particular type of character trait will require more text than this chapter allows.

Fear

From all the information above it would seem that we live in a world that is controlled by fear and to a degree this is the case, we are fearful today of so many things, we fear the sun, diseases, bad health, violence is everywhere, we do not let our children play freely due to the risk of predators, the risk of terrorism is ever-present, the list goes on and on, the fact is, that nothing creates a feeling more powerful within our minds than the risk of fear. Is it true, are we now living in the most dangerous times within our evolutionary history? Facts would argue otherwise, our life expectancy has increased generation on generation and statistically fewer people die today than 100 years ago.

What we do have now is instant communication across the globe, when a horrendous event occurs such as the September 11 terrorist attack in New York, we are instantly dialed into the event and those survival mechanisms within our brain are triggered, we become fearful. This subject is not one that we like to dwell on long, but we must remember that the bully can be overcome; there are ways to combat their effects.

Bullying and the effects that this behaviour creates is a very serious issue, if they are not addressed than the psychological scars will continue throughout life. We will never eradicate this behaviour completely, acknowledging this gives us the tools to understand the behaviour, identify those that carry out the bullying and their victims, this in turn allows for real control and management for both parties. They are both victims, the bully, from exposure to today's accepted cultural behaviour patterns, the lack of correct parental guidance, a lack of nurture when a parent is lacking the child's best interest at heart and from a left over survival mechanism within the brain, that two thousand years ago did not have the moral ties to bind this behaviour and the individual who is subjected to the bullies behaviour, who is also a victim of culture, lack of understanding, support, love, nurture and many more. My hope is that this information will at least help a little, if it does, then at least a few will grow up without the foreboding baggage given to them by a bully and today's society. The real message here is that the effects of childhood bullying on the mind can be changed, the mind has the ability to rewire it's neural pathways, plasticity in the brain can occur, especially in the early years, but the process needed to achieve this takes understanding and ultimately a willingness to recognise and act.

References

Alpern, S, B. (1998) The Women Warriors of Dahomey. Published by C. Hurst & Co Ltd. London.

Anderson, S and Sovik, R. (2000). Yoga: mastering the basics. Published by Himalayan Institute.

Asken, M, J. PhD & Grossman, D Lt. Warrior's mindset (2010) Warrior science publications.

Answers (2013). How many breaths do humans take in an average lifetime. Accessed on 01/10/2013 @ http://wiki.answers.com/Q/How_many_breaths_do_humans_take_in_a_average_lifetime

Bargh, J, A. (2006) European Journal of Social Psychology Eur. J. Soc. Psychol. 36, 147–168 (2006) Published online in Wiley Inter Science (www.interscience.wiley.com). DOI: 10.1002/ejsp.336 John Wiley & Sons, Ltd.

Bandura, A. (1971) Social Learning Theory. Ibrary of Congress Catalog Card Number 75-170398. Published General Learning Press.

Cartledge, P. (2007) Another view. Professor of Greek History on 300. Accessed on 15-12-2013 @ http://www.theguardian.com/film/2007/apr/02/features.arts

Chang, C, W, C. Brent, L, J, N. Adams, G, K. Klein, J, T. Pearson, J, M. Watson, K, K. and Platt, M, L. (2013) Neuroethology of primate social behavior. Edited by John C. Avise, University of California.

Cannon, W, B. (1915)Bodily Changes in Pain, Hunger, Fear and Rage. New York, NY:D. Appleton & Company;1915.

Chapél R, phd, (1991) course book S-101 V-9.9.8

Cherry, K. (213). About.Com Psychology. Social and Emotional Development in Early Childhood. Accessed on 31-07-2013 @ http://psychology.about.com/od/early-childdevelopment/a/Social-And-Emotional-Development-In-Early-Childhood.htm

Cleary, T. (2009) Training the Samurai Mind. Published Boston & London.

Corbetta, M. (1998) Frontoparietal cortical networks for directing attention and the eye to visual locations: independent or overlapping neural systems? Proc. Natl. Acad Sci. USA, Vol. 95, pp. 831 – 838, Febuary 1998 Colloquium paper.

Cohen, N. Poldrack, R. Eichenbaum (1997) Memory for items and Memory for relations in the Procedural/Declarative memory framework. Psychology press, an imprint of Erlbaum (UK) Taylor & Francis Ltd.

Cooper J, (1989) "principles of personal defence" Paladin Press

Czartoryski, A. (2011) Amazing Hunter-Gatherer Societie still in existence. HunterCourse.Com. Accessed on 02-05-2014 @ http://www.huntercourse.com/blog/2011/05/amazing-hunter-gatherer-societies-still-in-existence/

Darryl W. Schneider, John R. Anderson Cogn Psychol. Author manuscript; available in PMC 2012 May 1. Published in final edited form as: Cogn Psychol. 2011 May 1; 62(3): 193–222. doi: 10.1016/j.cogpsych.2010.11.001

Dawkins, R. (1976) The Selfish Gene. Publishers, Oxford University Press.

Department of Education (2013). Preventing and tackling bullying Advice for headteachers, staff and governing bodies. Document available @ www.education.gov.uk Crown copywrite (2013).

Diamond, J. (1997) Guns, Germs, and Steel. Published by Random House, 2013.

Divine, M. (2010) Bullying Hurts: How bullying takes our brain's ability to adapt and turns it against us. Posted on 07 September 2010 by admin. Accessed on 04/08/2013 Available @ http://www.michaeldevinecounseling.com/blog/bullying-hurts-how-bullying-takes-our-brains-ability-to-adapt-and-turns-it-against-us

Dougherty, p. Ph.D. The department of Anesthesiology, pain and medicine. At MD Anderson cancer centre. Chapter 2 Somatosensory Systems. The department of neurobiology and anatomy at The University of Texas medical school at Huston. Published on Neuroscience Online Accessed on 04/03/2014 @ http://neuroscience.uth.tmc.edu/s2/chapter02.html

Due, Pernille, Dr. (2005). bullying and symptoms among school-aged children:international comparative cross section study in 28 counties. European journal of public heath. accessed on 28/07/2013 @ http://eurpub.oxfordjournals.org/content/15/2/128.short

Duhigg, D. (2013). The Power of Habit: Why We Do What We Do in Life and Business. Published Random House Books.

Evans, J, St, B, T, (2006) Dual System Theories of Cognition. Centre for Thinking and Language, School of Psychology, University of Plymouth, Plymouth PL4 8AA, UK accessed on 20/07/2013 link, http://csjarchive.cogsci.rpi.edu/proceedings/2006/docs/p202.pdf

Espelage, D, L. and Holt, M, K. (2001) Bullying and Victimization During Early Adolescence: Peer Influences and Psychosocial Correlates. Howarth press, Inc.

Evens, J, St, B, T. Dual System Theories of Cognition: Some Issues. Centre for Thinking and Language, School of Psychology, University of Plymouth, Plymouth PL4 8AA, UK

Fernandez-Ballesteros, R. (2002) Encyclopedia of psychological assessment. Anxiety Assessment. Published by Sage.

Gardener, D. (2008). The science of fear. Published July 17th 2008 by Dutton Adult.

Ferrarelli, F. Smith, R. Dentico, D. Riedner, B, A. Zennig, C. Benca, R, M. Lutz, A. Davidson, R, J. & Tononi, G. Experience mindfulness meditators exhibit higher Parietal-Occipital EEG Gamma activity during NREM sleep. PLoS ONE 8(8): e73417. Doi:10.1371/journal.phone.0073417. published August 28-2013.

Gascoigne, B. (2001) History World. History of the Domestication of Animals. Accessed on 07-05-2014 @ http://www.historyworld.net/wrldhis/plaintexthistories.asp?historyid=ab57

Goodall, J. (2014). About Chimpanzees. The institute of Canada. Accessed on 10/03/2014 file://localhost/@ http/::www.janegoodall.ca:about-chimp-so-like-us.php - Intelligence

Goldstein, D, S. and Koplin I, J. (2007). Evolution of concepts of stress. Clinical Neurocardiology Section, National Institute of Neurological Disorders and Stroke, Bethesda, MD USA
(Received 4 December 2006; revised 15 January 2007; accepted 19 February 2007)

Grossman, D. Lt. (2004). On Combat: The Psychology and Physiology of deadly conflict in war and in peace. Millstadt, Il: PPCT research publications.

Grossman, D. Lt. (1995). On Killing: The Psychological cost of learning to kill in war and society. Publishers, little, Brown and Company.

Hardy, G. Great minds of the Eastern Intellectual Tradition. The Great Courses. University of North Carolina at Ashville, Yale University. Downloaded 2014. Publisher The Teaching company (2011).

Heim, C. Dietmar, S. and Eckhard, N. (2006). Towards an understanding of involetary firearms discharge – possible risks and implications for training. Policing an international journal of police strategies and management. Volume: 29 Issue: 3 Pages: 434-450 DOI: 10.1108/13639510610684683 Published: 2006.

Hobbes, T. (1660) Chapter - Of the natural condition of mankind as concerning their felicity and misery. The Leviathan. Accessed on 10-12-13 @ http://oregonstate.edu/instruct/phl302/texts/hobbes/leviathan-contents.html

Hogan, M. and Sherrow. PhD. (2011) The Origins of Bullying. Originally published in Scientific American, December, 2011

Holló and Novák: The manoeuvrability hypothesis to explain the maintenance of bilateral symmetry in animal evolution. Biology Direct 2012 7:22.

Huxley, E. (1985) Out of the Midday Sun. published Great Briton by Chatto & Windus Ltd.

Indian Country Today Media Network, LLC (2014) accessed on 24/02/2014 @ http://indiancountrytodaymedianetwork.com/2013/11/07/joseph-medicine-crow-last-plains-indian-war-chief-turns-100-152106

Invisible Laws. Sticky Decisions – Anchoring and Adjusting Posted on June 3, 2013 by Dan accessed on 20/07/2013 @ http://www.dangreller.com/sticky-decisions-anchoring-and-adjusting/

James, W. (1890). The principles of psychology. Autherised Dover edition published (1950), first published by Henry Holt & company (1890).

Jacobson, R. (2013). Scientific America MIND. Brief history of virtue. Vol. 24. No 5. November/December 2013. Published bimonthly by Scientific America.

Jefferys, W H. and Berger, j O. (1992) Ockham's razor and Bayisean analysis. American Scientist. Vol. 80. No 1 (January-February 1992), pp. 64-72. Published by Sigma Xi, The Scientific Research Society.

Jerath, R. (2006). Paranyama breathing. Published online at PubMed.gov. Accessed on 01/10/2013 @ http://www.ncbi.nlm.nih.gov/pubmed/16624497

John F. Kennedy Presidential Library and Museum (2014). The Cuban Missile Crisses. Columbia Point, Boston MA. Accessed on 07-03-2014 @ http://www.jfklibrary.org/JFK/JFK-in-History/Cuban-Missile-Crisis.aspx

Kahneman, D. (2011). Thinking Fast and slow. Penguin books Ltd.

Kosinski, R, J. (2010) A Literature review on Reaction Time. Updated September 2013,. Accessed on 17-02-2014 @ http://biae.clemson.edu/bpc/bp/lab/110/reaction.htm
Kurtz, L, R. and Urpin, J, E. (1999). The Encyc;opedia of Violence, Peace and Conflict. Academic Press.

Lee, B. (1975) Tao of Jeet Kune Do. Ohara publications incorporated. Linda Lee, all rights researved (1975).

Lee, J, W. (2005) Focus on Gender Identity. **Tinbergern's** 4 questions. © Nova Science Publishers, Inc. New York, 2005

Liepert, J. MD. Bauder, H. Ph.D. Miltner, H. Ph.D. Taub, E. Ph.D. and Weiller, C. MD. (2000) Treatment-induced cortical reorganization after stroke in humans. Copyright 2000 by American heart Association.
Lorenz, K. (1963). On Aggression. Translated by Wilson, M, K. Edition (1966), (2002), Published by Routledge.

Meichenbaum, D. (1976). Psychology Wiki. Stress inoculation training. Accessed on 11/07/2013, at http://psychology.wikia.com/wiki/Stress_Inoculation_Training

McLeod, S. A. (2007). Simply Psychology. Pavlov dog's. Accessed on 16/06/2013 @ http://www.simplypsychology.org/pavlov.html

Mickevicien, D. Motiejunaite, K. Karanauskiene, D. Skurvydas, A. Vizbaraite, D. Krutulyte, G. and Rimdeikiene, I. (2001), Gender-Dependent Bimanual Task Performance. Published Medicina (Kaunas) 2011;47(9):497-503.

Murray, K, R. (2004) Training at the speed of life. Vol, 1. The definitive textbook for military and law enforcement reality based training. Armiger publications, Inc.

Nadkarni, N, A. and Deshmukhi, S, S. (2012) Mirror movements Annals of Indian Academy of Neurology, January-March 2012, Vol 15, Issue 1.

National Geographic Society (2014) The Development of Agriculture. Accessed on 07-05-2014 @ https://genographic.nationalgeographic.com/development-of-agriculture/

National Womens History Musem (2013). Ann Hennis Trotter Bailey (1742-1825). Accessed on 05-12-2013 http://www.nwhm.org/

Native Words – Native Warriors (2007) Smithsonian Institute, National Museum of the American Indian. Accessed on 19-12-2013 @ http://nmai.si.edu/education/codetalkers/html/chapter4.html

Newell, Allen and Rosenbloom, Paul S., "Mechanisms of skill acquisition and the law of practice" (1980). Computer Science Department. Paper 2387. http://repository.cmu.edu/compsci/2387
NOVA Online (1997). Amazing Heart Facts. Accessed on 01/10/2013 @ http://www.pbs.org/wgbh/nova/heart/heartfacts.html

NSPCC Inform. Statistics on bullying. Website accessed on 26-08-2013 @ http://www.nspcc.org.uk/inform/resourcesforprofessionals/bullying/bullying_statistics_wda85732.html

Newitz, A. (2008). brain scans reveal that teen bullies get pleasure from your pain. Accessed on 05-08-2013 @ http://io9.com/5079234/brain-scans-reveal-that-teen-bullies-get-pleasure-from-your-pain

O'Neal, G. (2013). American warrior. The true story of a legendary ranger. Thomas Dunne books. ST martin's Press, New York.
Peterson, P. and Seligman, M, E, P. (2004) Character Strengths and Virtues. A handbook and classification. Published by The American phsychological association and Oxford Press (2004).

Peterson, P. and Seligman, M, E, P. (2004 – 2013) VIA Institute of Character. Accessed on 27-12-2013 @ http://www.viacharacter.org/viainstitute/classification.aspx

Pinker, S. (2012) The Better Angles of our Nature. Why Violence has Declined. Published by Penguin Books 2012.

Pulford,B,D. B.Sc (1996) Overconfidence in human judgment. Thesis submitted for the degree of Doctor of Philosophy at the University of Leicester

Raine, A. (2013) The Anatomy of Violence. The Biological roots of crime. Published by Random House of Canada Ltd, Teronto.

Saitoti, T, O. (1986) Originally "My Circumcision" from The Worlds of a Maasai Warrior, pp. 66–76. Reprinted by permission of Random House, Inc.

Schrödinger, E. Source of he mind. Liberation of human mind. Published (June 16 2011) accessed on 16/06/2013. http://sourceoforigin.com/tag/erwin-schrodinger-sayings/

Schwartz, J. M.D. and Begley, S. (2002), The Mind and The Brain. Neuroplasticity and the power of mental force. Regan Books, an imprint of Harper Collins Publishers.

Shimmura, T. Eguchi, Y. Uetake, K. and Tanaka, T. (2007). Differences of behavior, use of resources and physical conditions between dominant and subordinate hens in furnished cages. Graduate School of Veterinary Medicine, Azabu University, Sagamihara, Japan. Published Animal Science journal 2007.

Silva, C. Cid, L. Ferreira, D. and Marques, A. (2011) Attention and Reaction time in Shotokan Athletes. Published Revista de Artes Marciales Asiaticas (2011), vol, 6 issue 1, p141 16p. accessed on 17-02-2014 @ http://eds.a.ebscohost.com.libezproxy.open.ac.uk/eds/detail?vid=6&sid=389cb1f5-4638-440e-93a6-9a977afa7678%40sessionmgr4003&hid=4203&bdata=JnNpdGU9ZWRzLWxpdmUmc2NvcGU9c2l0ZQ== - db=s3h&AN=62829617

Skulking, J. (2004). Adaptive Fear, Allostasis, and the Pathology of Anxiety and Depression. In Allostasis, Homeostasis, and the cost of physical adaptation. Published by the press syndicate or the university of Cambridge.

Smalley, S, L. Ph.D. and Winston, D. (2010). Fully present. The science, Art and Practice of Mindfulness. Published July 13th 2010 by Da Capo Lifelong Books.

Smith, J, B. and Alloway, K, D. (2013) Rat whisker motor cortex is subdivided into sensory-input and motor-output areas. Front. Neural Circuits doi: 10.3389/fncir.2013.00004. Published on 28 Jan 2013.

Solzhenitsyn, A, I. (1973). The Gulag Archipelago: 1918-1956. An Experiment in Literary Investigation. Published by Harper and Row Publishers Inc. (1974).

Stanford School of Medicine (2013). Number 2, Gait Abnormalities Accessed on 14/11/2013 @ http://stanfordmedicine25.stanford.edu/the25/gait.html

Stroke, American Heart association (2013) Treatment-Induced Cortical Reorganization After Stroke in Humans. Liepert, J. MD. Bauder, H. PhD. Miltner, W, H, R, PhD. Taub, E, PhD; Cornelius Weiller, C, MD. Accessed @ http://stroke.ahajournals.org/content/31/6/1210.long

The History Learning Site (2014) Why do people commit crime? Accessed on 24/04/2014 @ http://www.historylearningsite.co.uk/why_do_people_commit_crime.htm

Tyldesley, B. and Grieve, J, I. (1996). Muscles, nerves and movement, kinesiology in daily living. Published by Blackwell science Ltd.

TIBCO BLOG (2013). The editorial staff. Accessed on 25-09-2013 @ http://www.tibco.com/blog/2013/07/11/john-boyd-the-ooda-loop-and-near-real-time-analytics/

Tulving, E. (1985), How many memory systems are there? American psychologist, vol. 40, April 1985. Printed in USA.

Uttner, I. Kraft, E. Nowark, D, A. Muller, F. Phillipp, J. Zierdt, A. and Hermsdorfer, J. (2007). Mirror Movements and the Role of Handedness: Isometric Grip Forces Changes. Motor Control, 2007, 11, 16-28 © Human Kinetics, Inc.

Utterly Global (2013). Stand up, speak out end bullying. Accessed on 28-08-2013 @ http://antibullyingprograms.org/Statistics.html

Voltaire (1694 – 1778) The quotations page. Accessed on 24/04/2014 @ http://www.quotationspage.com/quote/26816.html

Wright, R. (1994) Moral Animal. Why we are the way we are: The new science of evolutionary psychology. Pantheon books, a division of Random house, Inc. New York, in 1994.
Wikipedia (2013). Amygdala. Accessed on 11/09/2013 @ http://en.wikipedia.org/wiki/Instinct

www.ingramcontent.com/pod-product-compliance
Lightning Source LLC
Chambersburg PA
CBHW081057220326
41598CB00038B/7137